Oak Island
and the
Real Treasure

It's Not
What You Think!

Lee Larimore

Copyright © 2022 by Lee Larimore

All rights reserved. No part of this book may be reprinted, republished, reproduced, or trasmitted in any way without the written permission of the author.

ISBN 978-1-62806-360-8 (print | paperback)

Library of Congress Control Number 2022917790

Published by Salt Water Media
29 Broad Street, Suite 104
Berlin, MD 21811
www.saltwatermedia.com

The background image of the maps on the cover is used courtesy of unsplash.com user Ruthie.

Oak Island and the Real Treasure

It's Not What You Think!

Contents

Oak Island Introduction i
Island of Mystery! 1
The Beginning of the Mystery 7
The Dig Begins 19
Tragedy Strikes! 29
The Curse of Death Again! 35
A Big Dig 47
A New Era Begins 51
Who Did It? 71
The Suspects 83
The French & English Military 121
The Templar Suspects 129
Knights Templar & The Ark of the Covenant 145
Ralph de Sudeley: Part 1 165
Knights Templar & Ralph de Sudeley: Part 2 179
So What is the Answer? 191
Symbolic Clues Leading to the Answer! 197
Important Things to Consider 219
Nolan's Cross 249
Nolan's Cross - A Map? 259
The Cross Uncloaked! 263
The Sacred Symbol Connection! 305
References 321
About the Author 329

Acknowledgements

First, I must thank my wife, Dora, for being so forgiving of all the time I spent working on this book.

I would like to express my sincere gratitude to the late Sir Henry Lincoln for uncovering the hidden secret that has opened the pages to an important story for history. I am indebted to Allysha Lavino, a dear friend to Sir Lincoln. She was an asset as a liaison as well as an author in her own right.

Also, many thanks to the late Fred Nolan for having the drive to find what others had ignored.

I have a deep appreciation to all who have come before in their challenge to solve the Oak Island mystery.

I am thankful for the Lagina brothers, Rick and Marty, who today still continue their quest.

Many kudos to the team at Salt Water Media for their patience and professional guidance.

Finally, the biggest praise is reserved for God in the knowledge he gave me and in his light to see the way!

- Lee Larimore
November 2022

Oak Island Introduction

Are you the type of person who enjoys a mystery that's like looking into a dark abyss, one that keeps you wanting to stare deeper within to find its secret? Maybe you enjoy a story that's like a spider's web that captures the reader inside its many strands, all the while hoping that you can solve the entrapment of its maze of mysteries. If so, then there is an ever-evolving mystery about an island off the northeast coast of Canada that can capture your curiosity to its fullest. That island is called Oak Island, which is among a group of three hundred or more islands located in an area of water known as Mahone Bay near Halifax, Nova Scotia. Not only is there an age old mystery of what some believe has been buried on this odd shaped island, but the forever question as to who put whatever it is there, when did it happen, and why. Another part of the mystery involves how to actually retrieve what is believed to be buried deep down in a pit within the earth, and if it can even be accomplished. So far, all of the attempts made during the past two centuries by the many treasure hunters trying to unravel the myriad of challenges and mysteries encountered on this island, have only resulted in an endless history of failures. The gripping lure of becoming filthy rich by finding an ancient treasure that could be worth millions of dollars in shiny gold coins along with piles of heavy silver ingots, or great works of invaluable writings by those of great historical prominence, has so far produced only a formidable quest with success just an elusive reward.

The history of this island began with the echoed tales about something that was imagined to be of an immense treasure buried deep in an underground pit, put there by pirate suspects of lore like those known as the infamous William Kidd or Blackbeard. There is even a suspicion there may be some of wise old King Solomon's ancient treasure buried deep underground brought there by the holy crusaders known as the Knights Templar. The island may even have a buried sailing ship from an unknown origin that is laden with treasure of gold and silver that time has sunken deep within an area known only as "The Swamp." Who would do that and why sink a ship there? There is also some huge boulders that are precisely laid out in the design of a Latin cross spanning an area covering several acres of land on the island. What was its purpose? Was it supposed to be a sign? But a sign for what? Six men have already met their death while searching for the suspected treasure with yet one more soul doomed to perish on the island as proclaimed by a rumored curse. A curse that supposedly was invoked by the death of drowning slaves trapped deep in the search pit they had just dug!

Will the curse come true? Strange happenings have always haunted this place: stories of ghosts dressed like British soldiers marching through the woods, a red eyed dog like creature that stalks anyone daring to be present alone. Unholy screams and howls that can be heard in the still of night, and even voices that are said to be so haunting as to make your skin tingle with chills. There have been unexplained accidents and untimely setbacks that were always happening to the crews and their equipment that was used in attempting to discover what treasure may be hidden.

There are so many questions awaiting an answer about what has happened on this island, and none have yet to offer any glimmer of hope toward an easy solution. Even though many have tried for more than two centuries to unravel the mysteries that yet remain, no one is any closer now than when those efforts first begun. Do you want to try?

The mystery was born a little more than 200 years ago, in the year of 1795 when some curious young lads discovered what they thought to be the location of a buried pirate's treasure, possibly that of a pirate like William Kidd. If it was treasure, it could make them wealthy and famous beyond their wildest dreams. But, for all of those who have attempted in the past centuries to discover what it is, Oak Island has only remained an elusive prize yet to yield an immense treasure trove of wealth. There have been numerous stories of what is there, none of them may be true, only time will tell. The mystery, or should I say, the many mysteries that shadow Oak Island have become some of the most perplexing in modern history in yielding any solutions toward unraveling their secrets. The more time that passes in an attempt to answer the mystery of what Oak Island is about, the more the hopeful theories come forth of who, what, why, and when. It is the mystique of the numerous old tales that have been echoed so often from days of old that there may be millions of dollars in gold and silver buried on the island secreted away by any number of a variety of suspects. Some believe a treasure was buried a few hundred years ago, others believe it could have happened even centuries before that, yet not even history has been helpful to confirm when. Many have thought that it was pirates, those legendary evil scoundrels who plundered the seas and cached their booty to probably someday retrieve, maybe it was thought to be their retirement stash. It may be as some have said that what lies hidden deep in this mysterious pit is that of an ancient find long ago excavated from beneath Solomon's Temple. Could it possibly be the most sacred of all – the Holy Grail?

The imaginations and dreams of countless men and women who have become involved in the mystique of Oak Island, have created in their own minds and that of others around the world, that whatever Oak Island represents today, is probably as great a mystery as time has ever witnessed! Ever!

There have been scores of mysteries of all types that have been echoed throughout history, some have been solved, but many still remain. There are still questions about if someone else may have been involved in the assassination of the U. S. President John F. Kennedy. What really did happen to Jimmy Hoffa, the teamster union boss whose body has never been found? What is beyond our universe, or is death really the end? Oak Island has been, and still today, remains a mystery not unlike those, but can it be solved?

Oak Island has at sometimes given up a small offering toward success, but usually ends up just yielding another teasing glimmer of false hope toward answering one of its many questions. But yet, the challenge continues, even as many have failed, others would begin their own relentless quest with aspirations of success. There have been millions upon millions of dollars that have been invested by any number of investors into the many search attempts that were begun over two centuries ago. Some searchers have even employed the use of heavy machinery to move away many tons of earth, creating massive deep and darkened perilous pits. Men have labored tirelessly until days ends, to dig shafts deep into the earth in attempting to discover the supposed treasure that may lie more than 200 feet deep below the earth's surface.

All of the different individuals and groups of treasure seekers that keep coming to the island in search of that elusive goal of finding treasure, are left to ask themselves the same question: will they ever become successful?

Oak Island continues to present a myriad of unending questions of which human history has played a very intriguing role as to the question of who really is involved in this enigma? Who was the creator of what is found on Oak Island? There are only deep and forever puzzling questions ever so entrenched in this very complex mystery!

The theories have been numerous ranging from whatever the mind can imagine as to what there is to be found, and to who actually did what, why, or when, yet not a single theory that has been presented so far can unequivocally answer any of those questions with 100% certainty. They are not even close!

The Island, as some would say, has been ravaged by the constant deep probing of searches into its underground that have occurred during the many decades of exploration. The numerous excavations, bulldozing, and the tunneling of the earth that has occurred on this odd shaped island has even caused some to condemn those who do the work as being destructionist! Parts of this island have even suffered from historical elimination and geological changes that were recklessly caused by some in their efforts to gain access to the elusive treasure of the "Money Pit." It makes one wonder, can all the continual probing of the area leave any treasure still intact? Has it been destroyed by what some have said has been nothing short of the raping of the area where the treasure was thought to be? Some searchers have quit either because they ran out of money in attempting their fruitless venture, or had their government permits expire, or maybe even suffered some kind of a fatal mishap. Unfortunately for many others, they probably just became overwhelmed by frustration and disappointment in not being successful, so they would just quit and humbly abandon Oak Island yielding to defeat!

Why have so many tried for so many years to answer the question of what Oak Island is all about? Is it really something there that is so important that men are willing to risk everything they have, even their lives? Haven't they known that those that came before them were not successful? Of course they did, but that is the ultimate challenge, to succeed where others have failed.

We have found the *Titanic* where she lies in the great depths of her cold ocean grave. Landed a manned rocket on the moon that enabled the astronauts aboard to walk on the moons surface, and then return successfully back to earth. Science has discovered the genetic DNA code of the human makeup and opened our eyes as to how living things are structured. If we try to succeed at a given challenge, a solution can be achieved. Some may take longer to accomplish than others, but positive results are achievable. Can we answer what we need to discover about Oak Island? Is it pirate gold that's buried beneath the surface, maybe precious gems also, or could it be as some have speculated, yet another form of treasure hidden long ago from us that's not meant to be found?

Some have rumored that it could possibly be the Ark of the Covenant, which carried the Ten Commandments, or maybe other precious religious artifacts, even the Holy Grail? Did the poor holy warriors of the religious crusades from hundreds of years ago aptly known as the Knights Templar come into possession of those holy items after finding them in Solomon's Temple? Did the Knights Templar bring those treasures to such a far away haven as Nova Scotia in order to evade the wrath of the Muslim empire who would destroy any evidence of western Christian belief? Are the Templar the constructors of Nolan's Cross?

Whatever Oak Island is clutching in its depths, whatever may someday finally be given as its answer, the world awaits. It is speculated that historians will hopefully get an answer soon to this long enduring mystery. There are some people who already have strong opinions about what Oak Island hides and maybe by whom, though none actually know. What do they believe? What will you believe after reading this book?

In writing this book, I hope that you will understand that reading all of the chapters as being necessary to fully grasp what an incredible journey this mystery has evolved into, and still continues. From the very first discovery of a possible buried treasure site on an island in North America in 1795, to the speculation of the Knights Templar being involved, to a deep suspicion of it being something even more mysterious that possibly has an invaluable religious connection. It will be intriguing! Every chapter in this book contributes its own essential piece toward understanding the complete story. I will try to capture the most relevant about each, some are more detailed than others, but knowing about them all, helps the reader to understand why each contributes to the final and complete understanding of what has created the mystery. When the Oak Island mysteries are finally solved, the participation by all who have been involved, whether in the past or in the present day, will hopefully receive the recognition that they deserve. No matter the depth of commitment in whatever degree each of their parts were toward solving this mystery, all will finally be a contribution justified, no matter how small their involvement.

Only when all of these pieces are fitted together like a completed puzzle, will the story picture be clear, for in this book there are the necessary parts which you will receive in order to fully understand and appreciate what an amazing journey it is to venture through history in solving the Oak Island Mystery!

If you have ever watched the TV series *The Curse of Oak Island* which is aired on the History Channel, you will enjoy this book as it lightly covers the chronology of the story that eventually answers all! If this is the beginning of your journey into this mystery, you will learn the story that has captivated the curiosity of others for centuries!

There are answers written herein that are intended to finally unmask the truth about what is the real secret of Oak Island! It is time to begin the search toward that answer! Are you ready?

Island of Mystery!

As I rolled over in bed, thoughts of what to do for the day became focused on being occupied at the computer doing more research about the activities of pirates during the Golden Era of Piracy. Sometimes as you lay in bed half-awake your best thoughts seem to formulate as you lie there and hope not to forget them! I needed to put some thoughts on the computer screen before they vanished. I had been trying to determine if any pirates, or pirate had ever been involved in a large scale adventure similar to what had occurred on Oak Island. Such an endeavor like that taking place anywhere on the coastlines along the Americas, or even in Europe, would surely have been known, or so I thought. What kind of protection for a treasure is so necessary that it would require an absolute commitment to stashing it in a location that was designed to just about be impossible for anyone to ever recover! Usually, as it has been recorded in history, pirates don't waste too much time nor energy to bury their treasures to any great extent as that which has been done on Oak Island. I had become perplexed about why such a place that has been rumored to have been done by pirates, also has been described as being so ingeniously designed, and intentionally foreboding in its construction, that it is literally a deadly venture to pursue. So far none of what I had researched was even indicating a single pirate possibly having conducted any such complex construction that has been determined to be like a labyrinth of underground booby trapped tunnels hundreds of feet below the earth's surface. There definitely was no record of any such venture, but reasoning would eliminate any record of something not wanting to be known. The whole idea that pirates could also be so secretive and knowledgeable about how to accomplish such an engineering feat, was a little hard for me to imagine from the beginning when I first knew of this mysterious story.

As the early morning hours had passed, with breakfast finished, I was back at the computer searching the various websites that referenced the histories of pirates. None of my searches so far had revealed anything which could give me the answer to what I was seeking. It seems that sometimes you just keep hitting dead ends when you need to find what should be an easy answer. There was one sizeable discovery: I didn't realize that there were so many pirates that had sailed the southern waters of the Caribbean who also ventured northward to the colder lands near Nova Scotia. History is replete with hundreds of names of individuals who either became pirates or privateers who had sought to plunder bounty on the open seas from the Atlantic to the Indian Ocean and beyond. I believed it was time to actually begin to think that whatever was buried on this island was so significant that it was going to require a lot more research into the stories about any possible suspects. I began to accept the idea that what had been originally done on that island in Nova Scotia occurring centuries ago, could only have been accomplished by the efforts of a very well organized group. It would be safe to speculate that it would probably require the size of a small army consisting of devotees who were sworn to absolute secrecy, and would surely have to endure an unknown number of weeks or months of strenuous labor.

What type of pirate, or group would be so committed on hiding something that was so ingeniously protected that it still remains elusive today even after centuries of efforts to uncover its secret? We can only imagine as so many others have that a unique and priceless treasure must be awaiting. But who would make it so difficult and dangerous to prevent its discovery? Why do such a thing?

As I sat there pondering what to search next, I slowly glanced at the many books I had accumulated that were scattered around the floor and on top of a cluttered desk. All the different books were beginning to number like a small library. There were of course books about pirates and privateers even a few books about Oak Island, Solomon's Temple, the Freemasons, Knights Templar, and a Bible containing the Old and New Testaments. There was even a large scale map of France, also an eight-foot historical timeline graph full of important dates made out of sections of poster board paper that was laid end to end. There were several loose leaf and spiral ring binders filled with hundreds of notes I had written as each book and article were read. I had become overly captivated with trying to uncover the mystery about that darn treasure supposedly buried 200 feet down in the earth on a 144-acre island in Nova Scotia.

Could the legend of Oak Island in Nova Scotia really have a hidden treasure? Ever since 1795 when Daniel McGinnis and a couple of his friends found a depression in the ground on that odd-shaped deserted island, it seems that a lot of people have concluded it must be worth the risk in trying to discover what is there. The cost of that risk though has unfortunately caused many to lose much, some have lost all they had, and to date in 2022, there have been six others who have given the ultimate in losing their lives!

The TV series, *The Curse of Oak Island* presented by the History Channel began airing on January 5, 2014 with two brothers deciding to join in the search efforts that others had long pursued in a quest to solve the mystery. That mystery had begun over two centuries before in 1795. The brothers, Rick and Marty Lagina from Michigan, had become interested in Oak Island during their youthful years after big brother Rick had read an article about the island which was published in the *Reader's Digest* in 1965. Little did he and his brother Marty, both now grown men, imagine that early in their lives the fantasy of them trying to find the treasure would actually become a reality. Today, in 2022, they both have teamed up with the History Channel in a very aggressive effort toward solving the mystery which has shadowed the island for centuries. So far they have found many very interesting artifacts that date back in time before Oak Island was even formally recognized by the government of Nova Scotia. Some finds possibly have dated as far back as the 13th century or sooner. Their attempts at excavating the so named Money Pit that others began back in the early 1800s, has yet to yield the unknown treasure of wealth. They have accomplished some success by continually amassing clues toward an understanding of what has been deemed a very profound and entangled historic mystery, yet not forgetting that it could yield a fantastic treasure!

I had recalled reading the *Reader's Digest* story myself back when I was probably in my early 20s, can't remember where it was, but I do remember reading it. I think back then what captured my interest was the fact that there was a page showing a sketch that depicted the Money Pit shaft that lead down to a treasure.

So here I was trying to put on paper any thoughts that may help in my understanding of what happened to make Oak Island into what it had become, a big mystery! When I first began to watch the television series about the Oak Island mystery, never did I expect it to be so complicated, complex, captivating and alluring! There could only be one approach to pursue in an effort toward finding a solution, and that was to critically delve into every aspect about every event and every suspects history that could give reason as to why the mystery even existed! It had to be an in-depth search into all of the history that is related to, or known about the mystery, then hopefully to assimilate everything into a hypothesis that could endure any critical test. Did I really think I could achieve that goal? Anyone wanting to solve a problem always hopes to find a solution to their endeavor, but that doesn't always happen, but you must try. In reading this book there is the hope that the reader wants to achieve an answer to the same thing I sought. If you have been following the TV series of Oak Island or reading a book or two about it, then by now you're hooked into also wanting the answer! Where we begin and where we end will ultimately measure the worth of the journey. Just like those who have come before in the search for the truth, it is time for us to take up the challenge!

The Beginning of the Mystery!

It was the unusual appearance of lantern lights on an uninhabited island near the coast of Nova Scotia late one spring night in 1795, that would capture the curiosity of a young man named Daniel McGinnis. What could his thoughts have been that night as he observed from afar what he assumed to be someone searching for something? I am sure his suspicious young mind was puzzled at what he saw. Could this be some thieves looking to steal something? But what was there to steal on this uninhabited wooded island that would require such robber-like actions? It has to be a buried treasure they search for!

That event has now become known around the world as the mystery of Oak Island.

What was witnessed that night was some type of unusual activity not known to occur on such an uninhabited island. It was the year of 1795, a time period when Nova Scotia was sparsely inhabited by permanent settlers. Nova Scotia is a peninsula located off the eastern coastline of Canada and northeast of the mainland U.S. It has a land surface area measuring about 345 miles long and 78 miles wide and is surrounded by the Atlantic Ocean. Nova Scotia, in comparison to other land masses, is about a third of the size of Florida, and is smaller in land mass than Ireland.

Nova Scotia had not yet been inhabited by the Europeans while the mainland of America had already been witnessing the beginnings of its settlements. Nova Scotia would populate later, with only a few small villages located distances apart, in a land that belonged to the Mi'kmak indigenous people of the First Nations. Nova Scotia would only begin its earliest settlement known as Port Royal in 1605. It was settled by a few French settlers located midway on the western side of Nova Scotia. It was one of the first European settlements this far north as most others had been established farther to the south on the mainland of North America. Another well known settlement that would come later was Halifax, founded in 1713, mostly inhabited by English settlers, located farther to the north on the opposite side of Nova Scotia and situated on a seaside hill. Halifax also had a protected natural harbor area with direct access to the open waters of the Atlantic Ocean, making it a desired location for use as a naval seaside port that would later become a major trading settlement.

McGinnis was from the area called Chester Provence, settled around 1759, which is about 40 miles to the south of Halifax. The village of Chester Harbor, a part of Chester Provence, was located about midway on the eastern side of the peninsula and was also the site of a sheltered sea port that had become home to many local fisherman. The population in the area consisted of only a few hundred residents who were fairly prosperous with nice homes surrounding Chester Harbor which leads out into Mahone Bay, the location of Oak Island. McGinnis was a member of a farming family that lived in the more rural area called Western Shore, which is about two miles south of Chester Harbor. Oak Island is located just offshore a few hundred yards from that farming area of Western Shore. So in my opinion, McGinnis was probably watching from his family homestead that evening where he could have easily looked across the open bay waters to see the lights on the island. There are so many differing thoughts as to the exact date or place of McGinnis's witnessing the beginning of the Oak Island mystery, so some speculation definitely must be accepted. We know enough that he saw something!

Why would there be lights on this uninhabited island during the hours of darkness? In the year 1795, oil lanterns or torches were the only primary source of lighting for homes, ships, or the movement of people at night. Maybe someone was looking for a place to hide something of value and did not want to be seen during the daylight hours. Maybe they thought no one would be watching this isolated, tree-covered island on a dark spring night. By using the cover of darkness their movements would likely not be seen, but what reason would someone have to be there? Could they be trying to hide something? Maybe they were looking for something that had already been hidden? This is the kind of suspicious activity often used by those trying to secretly recover something, perhaps a valuable pirate cache.

Mahone Bay, which has hundreds of islands, was an area where it was normal to see a local fisherman or sailing craft of some kind maneuvering around the nearby waters during daylight hours, but to do so at night was very unlikely during those times. Oak Island is recognizable as being one of the largest islands in the area with about 144 or so acres. It is also different in that it was the only island covered with oak trees at the time of McGinnis, thus the name Oak Island.

There were many old tales of pirates who had begun their infamous adventures upon the high seas of the Atlantic, ranging from the coastlines of Europe to Africa, and even later to the Americas. Their escapades had started in the early 1500s with activities steadily increasing well into the 1700s, with the years of 1650s to 1720s infamously being called the Golden Age of Piracy. Pirates sailed all of the oceans but were mostly found roaming the waters of the Caribbean, then northward along the Carolinas even journeying farther north toward Nova Scotia. Their mission was to maraud whatever would come in their way. The nationalities (or countries of origins that most pirates claimed to be from) were France, England, Spain, Portugal, the Netherlands, and Ireland. Their prey would be any merchant ship usually hailing from countries other than their own. It really didn't matter to some pirate crews if they even attacked ships from their own countries. Occasionally a coastal harbor would become a necessary port of call for needed repairs, or to replenish the ships depleted supplies. Sometimes an unsuspecting port community would unfortunately suffer unwanted consequences of being pillaged, or even worse by the acts of an uncivilized crew.

One pirate, who roamed the nearby East Coast of the Americas, was the infamous Blackbeard, an Englishman whose real name was Edward Teach. Blackbeard was easily identified by his long beard and his usual wearing of multiple pistols strapped across his chest. During a short period in his lifetime, Teach lived in Bath, North Carolina. Another infamous pirate was Captain William Kidd, (1654-23 May 1701) born of Scottish descent, who at one time had actually been an authorized sea-faring agent for the English government. The English government had empowered Kidd to hunt pirates who had become a threat to the English ships, but later would himself become a hunted pirate. There were others like Henry Morgan and Calico Jack, Anne Bonny who was the best known female pirate, and Peter Easton who sailed the waters around Newfoundland. It was a lengthy list in the beginning of the 16th century that would continue to grow until the 18th century.

Plenty of stories have been told of valuable treasures supposedly being buried in many different locations during those years when pirates freely roamed the oceans. There was always some old salt as they were called, a veteran sailor of age perhaps, who would be more than willing to spin a story or two to entertain those who would turn an ear. Mostly the stories were just hearsay, embellished surely when only told by word of mouth that would be repeated down through the decades. Maybe some of what was said did have just a hint of truth in those old tales. In the past, there had been an occasional sighting of a ship or two that was seen flying the skull and crossbones atop the main mast that did frequent these nearby coastal waters. The eerie flags were known to be the symbol of those that sailed not for a country, but for themselves, and were seen as a fearful warning to other ships that they could become the prey and the bounty.

Was someone possibly looking for a pirate treasure that night on Oak Island? Could any of those old stories really have some truth to them? Who knew what was really true in those days or was just spoken to fabricate something for easy entertainment. In those days there were too many tales to discern the actual truth from ones of mere imagination.

Many accounts have McGinnis going to the island by boat soon after that night to investigate the sighting of those mysterious lights. I wonder what his thoughts were as he crossed Mahone Bay from the mainland to the island just a few hundred yards away. It is not for sure exactly what his age was at that time. Some indications were he was about 16 years old, others have claimed he was older, maybe even in his early 20s. His actual age is questionable, but whatever his age was, he probably already had begun thinking that he was about to go on an adventure that could lead to an unknown discovery.

After searching around the forested area of the island, which was about 144 acres in size and shaped like a baby elephant, or as a peanut as some have envisioned it, McGinnis came upon an odd sight of a block and tackle rig that had been hung by rope from an oak tree. The island was covered with oak trees, the only island in Mahone Bay to have that kind of tree growth. It was the unusual sight of a block and tackle rig hanging from one of those trees, located over a depressed cleared area of ground about 13 feet in diameter that captured his curiosity. The circular depression appeared to have resulted from the excavation of the soil that looked to have been done some years back in the past. The area he had discovered was a small open area surrounded by mature oak trees. The ground in the depressed area had the remains of some old tree stumps, surely indicating that someone had purposefully intended to clear that area in order to do some form of work. The ground was also showing signs of some new tree growth giving evidence that the site had not been recently altered. One can only imagine what was going through McGinnis's mind after finding this discovery. Probably his thoughts were of those old stories about buried treasure being cached by a pirate on an isolated island possibly being true, and thinking he just may have found such a treasure site!

Block and tackle rigs are used for either lifting or lowering of heavy objects; it was also standard equipment on sailing ships for the purpose of moving cargo on and off the decks, but also for the raising of the ship's sails. Why would the rigging be there? It was found hanging from the oak tree in a position to either lower a heavy chest of treasure below, or was left there after something had already been raised from the ground beneath? Who had used it? Somebody had been doing some unusual activity. Whatever the scenario was, it had the curious attention of McGinnis. Further examination was needed!

McGinnis probably knew that the best thing to do was to get some help and start digging. He would return to the island days later, bringing shovels, pick axes, and what he thought may be necessary for an excavation. This time McGinnis was accompanied by two other young men, John Smith, and Anthony Vaughn, all of whom were known to be friends and all were about the same age, as the story is told. It wasn't long after they had started their digging, maybe only removing a couple of feet of earth, when they hit some flat rocks that were laid in a pattern that covered what they thought would be the treasure. Upon removing the flat shale like rock, they only found more dirt beneath them to dig. The ground where the digging was being done was not as firmly packed as the surrounding undisturbed area. They dug deeper and began to notice evidence of previous shovel marks on the sides of the depression which now gave proof of a deepening pit. At 10 feet down they came upon some logs that had been laid like a flooring across the width of the dig area, which was about eight-feet in diameter. This had to be the access to the treasure below, or so they thought. Upon removal of those logs, there was still more easy digging of the dirt and still more signs of previous excavation work of what they now know is not just a pit, but a deeper shaft! After another 10 feet of digging down they came upon another layer of logs laid across the shaft, some of which were like the first logs, appearing to be very old with some rotting at the ends. Realizing that they now had dug down 20 feet into what was possibly becoming a dangerous situation for a cave in, and not having the right equipment to continue the dig, the decision was made to stop. They could only accomplish so much in attempting to hoist the dirt from the bottom of the shaft by using a pulley and bucket set up hanging from the oak tree above. No treasure, probably frustration, feelings of disappointment, lost time, lost money, they needed a new approach with the right equipment. It had become a bigger task then what they were prepared to attempt. They would have to quit for now, they would though come back, but not until some years later!

Pirates were known to bury their treasures usually in shallow pits, but the dig on Oak Island was definitely deeper than what old lore had told of the practice. The trio knew now that their only hope of successfully going any deeper into the shaft would require more manpower, equipment, time, and lots of money to do the job. But, they had to attend to their regular livelihoods of farming in order to provide for their families' survival. The only thing to do now was to search for someone willing to invest the funds needed to further their adventure. They had to be careful not to let the discovery be known to anyone whom they couldn't trust, or all could be lost before it was found. In an effort to conceal their find, the shaft was filled back in so as not to be evident to anyone nosy. Whoever had originally dug this shaft knew what they were doing, it had to require a tremendous amount of laborious work. There must be something of an extreme value to hide, otherwise why had so much work been done to bury something? Was it really a treasure of wealth that had been secretly hidden away sometime ago by the efforts of those yet unknown? Why did no one know of this existence on Oak Island?

In the years that followed, the trio's digging attempts were seemingly put on hold while hopefully awaiting to find an investor. Two of the trio would keep the find protected by purchasing some of the land on the island for themselves. In 1765, just 30 years earlier, the Nova Scotia Province was surveyed by a former British army officer named Charles Morris who was tasked to create maps of the Canadian maritime region. While on his mapping mission, Morris had mapped Oak Island dividing it into 32 lots each measuring approximately four acres in size. Morris had also mapped the layout of Halifax plus other major towns, eventually being appointed the surveyor general of Nova Scotia. John Smith would purchase lot #18 in the year 1795 near what would eventually be called the Money Pit and later would build a house for he and his eventual wife, Ann Floyd in 1799. Daniel McGinnis got married, bought property on the southwestern end of the island, and also built a house. Anthony Vaughn who also got married about this time period remained on the mainland. Vaughn's family had acquired a large tract of land in the Chester area through a homestead grant given to them by the government. Early settlers would be given land as an enticement to settle uninhabited areas like those around Chester. The Chester area had a nearby river which eventually was named after the Vaughn family. How the land purchases of the Island properties and the houses were funded by McGinnis and Smith, who were some young people known to be of modest means has always been viewed as suspicious. There has long been a suspicion that there was some treasure found on the island before any of the future major excavations had ever begun! Did the trio of McGinnis, Smith, and Vaughn actually find something that enabled them to better themselves above others of modest means who also lived in the same area? All of the tales about Oak Island being in an area of past pirate activity, especially with the most talked about pirate being William Kidd, may have been true about him burying a treasure of enormous wealth in the surrounding areas. In another chapter of this book, it will be covered in depth that there is a possibility that some treasure on Oak Island had already been found. Even if the trio did find some pirate loot of gold and gems, they didn't find the most valuable treasure! It is hopefully still there as some believe. But what is it?

Sometimes it just seems like time goes by quickly, yet it would take the trio of friends about seven to eight years before they would find an investor willing to chase what could be just another dead end dream of riches hidden beneath the ground. Between having to spend their time providing for their families, I am certain they were also spending some spare time searching around the island. Waiting for the passing of time for the trio to start a major excavation of the pit area can only be imagined to have seemed longer then they surely wanted to wait.

A person can feel the empathy for anyone suffering from the anticipation of obtaining something that can completely change their life and just within reach, yet escapes the grasp of possession. Just enduring the wait could make the experience seem painfully overwhelming. The trio of young men knew of the many tales of buried treasure that could be hidden around the lands of the New World, that were said to be anywhere up and down the east coast of the North American continent. Some of those tales were either believed to be mere fantasy as they were yet to be found, or were probably hidden without clue so as to never be found. Yet there are treasures to be found as we know what has already been discovered throughout history and even still today. There have always been those who seek treasure in hopes of finding that elusive buried treasure of golden loot and much much more! Many treasures have been said to be buried somewhere by the oceans' outlaws or others like them. Some treasure seekers will even scour the oceans bottoms searching for that sunken ship laden with tons of wealth which had become part of Davy Jones' Locker at the bottom of the sea. Any person who would risk their own fortune on the dream of what three young boys, now men, had described as the chance to uncover a hidden treasure on an island of hope, would indeed have to be a very adventurous individual. Perhaps another hopeful dreamer like them!

It was Anthony Vaughn who would eventually be successful in enticing a man by the name of Simeon Lynds to risk his money and time. Plus Lynds would also recruit several others that were willing to become investors in what they would call the Onslow Company.

Onslow is the name of a community about 60 miles north of Halifax and just south of another community called Truro which will be mentioned later. There are as with some Oak Island stories differences to how Lynds got involved, that give some accounts as Lynds being a merchant who happened to meet McGinnis in the Western Shore area located near Chester. Another story has Lynds as a doctor, who performed the delivery of a baby for one of the other trio's members' wives. However Lynds got involved, the adventure to uncover what was on Oak Island had begun, still continuing to this day. The group of investors was a list of very respected high profile people starting with the future speaker of the assembly for

Nova Scotia named Samuel George Wiliam Archibald. The future speaker also enlisted his nephew, Captain David Archibald, who was described as a farmer but may have been called captain from serving in the military. Also part of the group was the Sheriff of Pictou County that is located north of Halifax but near Truro, another investor named Thomas Harris, and one Colonel Robert Archibald, a government land surveyor who had mapped the community of Onslow. Colonel Archibald was also the local justice of the peace. Colonel Archibald and S. G. William Archibald were brothers. Anthony Vaughn was looking to find the right people and these were some pretty influential people for those times. They had the money and they certainly had the conections to get any legal documents needed or the ability to obtain any necessary governmental approval for permits.

Simeon Lynds had just given Vaughn, McGinnis, and Smith, the means to proceed with their long awaited pursuit of treasure that hopefully still laid beneath the surface of Oak Island. Their quest had been renewed!

The Dig Begins

This chapter is only one piece of the puzzle that must be acknowledged as it does have the potential in the end to answer some of the other mysteries that have created Oak Island! Yes there is more than one mystery shadowing the island! Only when whatever is at the bottom of the so named Money Pit is discovered, will the story even begin to be unraveled. Even then there may still be more questions needing answers to about what is found, or what is not found!

The big search began in June 1804 under the leadership of a man know as Colonel Archibald, who with a few other investors of the Onslow Company, would land their sailing ship on the shores of what is now known as Smith's Cove, Oak Island. The dig would begin slowly as the pit was now filled with debris, rain water, and mud that had accumulated over the years after the young trio had abandoned their precious efforts. Smith, McGinnis, and Vaughn would unite again with help from their neighbors, plus some others, as the dig began. This time, the searchers were equipped with better block and tackle rigging that had been provided by the investors. After clearing the shaft down to about 20 feet of what had originally been backfilled from the trios initial digging, the Onslow crew hit a log platform at about 30 feet down. Maybe this was the depth of the treasure, not so! After removing that layer of logs the digging would continue downward again, revealing even more evidence of much earlier shovel and pickaxe marks on the sidewalls of the shaft. The process of excavating and removing the soil in the pit continued with the same results each time finding more logs at 40, 50, 60, 70, and even farther downward to 80 feet! Who would bury something so deep? This had to be, as I am sure they supposedly also believed, something of such an incredible value as it needed to be so difficult for retrieval that it was critical to hide at such an unusual depth! I would also think that at this time in the search process the investors along with their crew began to realize that this was not going to be an easy nor inexpensive treasure hunt. The evidence left by the amount of original work done to create the pit could have only been possible by the efforts of a large labor force! Surely something was thought to be hidden very deep within the depths of this island! They had to believe it was something very extraordinary! Why else would some much work be done?

The dig was now down to 90 feet, but there was something different this time – it was a stone slab! This stone slab would lead to a very mysterious question: what was the purpose of this stone doing on Oak Island? The slab was about two feet long, 16 inches wide, and about 10 inches thick. One side of the slab that was facing up was smooth, but the underside when turned over had strange geometric shapes along with symbols possibly similar to those once found in ancient script or hieroglyphics! No one could understand what the strange symbols meant, but the inscription had to convey something toward understanding what the pit was, or what it concealed. One thing that was determined though, that the slab did not come form any area known to be around Nova Scotia! The slab was later determined to be some type of granite that possibly had come from the lands located around the northern areas of Europe. The interpretation of the markings on the slab are still a mystery to this day because it has since disappeared from known existence over the centuries. Some local history of the stone did reveal though at one time it was being used as part of John Smith's fireplace when he built his home on Oak Island. The stone was later reported to have become displayed as a novelty piece in a newspaper printing house located on the mainland, good for local conversation about the mystery, I assume. Some did try though to interpret what had at one time been copied from the stone before its disappearance. The markings had been interpreted into a couple of differing versions as follows: one claims it reads as "Ten feet below is two million pounds buried," another deciphering reads "Forty feet …" Experts have really never agreed on what it says, why it was there, or what it was meant to be. A mystery that someday might be solved, if it could ever be found!

After removing the stone at 90 feet, the work continued downward another few feet before the work stopped for the day. Before exiting the depths of the pit, a worker poked the ground with a long metal bar below where the 90 foot platform had been, striking a solid structure which they hopefully believed was the top of a treasure chest. The crew had finished their work for the day believing that there would only be a few more feet to excavate before finally retrieving the treasure. As the workers were preparing to be raised from the shaft they noticed there was a slight seepage of water beginning to enter the shaft from below, thinking that they could contend with that the next day. Unfortunately, upon returning the following morning the Onslow crew were surprised to find that the shaft had filled with water rising up from below to about the 65 foot level. Little did they realize this would be the start of what still continues to this day to be the nemesis of the digs in the Money Pit.

The efforts of that day now needed to be focused toward bailing out the water from the shaft, but the efforts became fruitless. It seemed that as fast as they would bail a few buckets of water out, the shaft would fill back up again to the same level as before. The investors knew their efforts were useless to try further, so they temporarily shut the operation down to re-group with another plan to empty the shaft. It was later that year when the Onslow crew returned, now equipped with the aid of a water pumping system in an effort to remove the water. Unfortunately, the pump lacked the suction needed to properly lift the water. The equipment was just not as good as was needed to do the job, failing almost immediately. It was now late in the summer and would soon be time for the local workers to return to their farms for the fall harvest. Colonel Archibald would call a halt to the operations for that year with plans to hopefully return the following spring.

The spring of 1805 did begin with the return of Colonel Archibald, who remained in charge of the Onslow operations, with a new plan to approach the treasure by creating a tunnel next to the original. This time, he would eventually attempt to tunnel under the bottom of the original shaft. The belief was that the treasure possibly laid beneath where they found the 90 foot slab, especially after the workers had hit what may have been a solid wooden treasure chest. By approaching the area from underneath at about 110 feet, the thought was they could avoid the water that was still in the original shaft. The plan seemed to be going well until the diggers got to within a few feet of the target location, that is when the sidewalls began to turn into wet clay and started to cause a cave in of the walls. Fearing for their lives the workers were forced to quickly flee the depths of the tunnel. Attempts were made to bail the water out from the new tunnel, but had the same results as experienced before with the shaft continuing to fill to the 65-foot depth. All of this was too costly in man-hours and supplies plus the failures were taking a toll on the investors dreams. The Onslow Company was done! After two years of investing most of their wealth, they all called it quits.

It would almost be fifty years later in the year of 1849 before another significant attempt would be made to seek treasure on Oak Island.

The next venturous group to try was the Truro Company, formed by a couple of the former Onslow investors to include Anthony Vaughn, now in his late sixties, Sheriff Thomas Harris, and a Dr. David Lynds who probably was a relative of Simeon Lynds. The other investors included a mining engineer, who was a mysterious figure named James Pitblado (more about him later) plus others who kept the historical records of the events and helped with the daily operations. Both of the previous digs had been filled in by John Smith leaving the original shaft needing to be cleared once more. After renewed digging of just a few feet down into the original shaft, the digging had to stop to remove the old broken water pump which had been left by the Onslow Company. After the pump was removed, the Truro crew continued to dig down for another 12 days stopping at about 86 feet where they found some previous wooden bracing of the pit walls. This is the work that had been done by the Onslow Company. The work was stopped on a Friday for the usual weekend break.

Returning that Monday, the crew was disappointed to find that the shaft had filled with water rising to about the 65 foot level just as it had done in the past with the Onslow Company. This time knowing that the efforts used in the past to remove the water had failed, a new approach to avoid the flood water was ordered by James Pitblado. In the past, Pitblado had worked with a drilling tool called a pod auger to search for coal, and he planned to use the pod auger to bore down into the shaft hoping to retrieve a sampling of what was below the 90 foot level. Pitblado believed that the treasure was just below where the 90 foot stone slab was found. The auger would drill down along the inside of the shaft stopping at five different locations to take samples from each drilling. At first the usual stuff was found on the drill point, mud, some various types of wooden pieces, presumably from the planks others had used to shore up the sides of the shaft, plus some kind of a plant fiber like that from the husk of a coconut. Coconut fiber was often used aboard sailing ships departing the southern Caribbean waters as an extra cushion to protect valuable cargo. The tropical climate of the Caribbean was a prime environment for the growth of the coconut tree which was not found elsewhere in the Americas. The usual practice of stowing cargo, such as flour, fruit, salt, cannon powder, and more was to store items inside of stacked wooden barrels that would then be secured with ropes for travel at sea. Other goods like stoneware, china, antiques, art works, bottled liquids like wine, or similar valuables may have been given special treatment that required some type of special protective cushioning. What would require this type of extra protection after being buried so far down in the ground?

The last pod auger probe of the five had a shiny item removed from the auger contents which was quickly pocketed by Pitblado. Jonathan McCully, who was one of the investors, had observed the removal of each of the drills findings and asked Pitblado what he had put in his pocket. Pitblado claimed it was not anything of value. Pitblado disappeared the next day after finding what was possibly a piece of the treasure, and he never returned to Oak Island! It is told though that Pitblado did show one man named Charles Archibald of Londonberry, Nova Scotia what he had found, and Archibald claimed that it appeared to be a gem stone, or a jewel!

Archibald is said to have come afterwards to see John Smith who owned the shaft area where the possible jewel was found, offering to buy all of his land holdings on Oak Island for many times what it was worth. What did Pitblado really show to Archibald? Yet, another mystery!

The Truro group continued their search with the digging of a third shaft next to the original Money Pit. The idea of the third shaft was to eventually create a water outflow tunnel that would drain the original shaft into the third. The crew was now digging down through some very hard impervious clay when they began to surmise that the water entering from the original shaft could not come from ground seepage but had to come in by some other means. As the diggers neared their target depth of around 110 feet they would begin to dig toward the original shaft. As the diggers got close to their target depth, the walls of the new shaft began to seep water into their shaft so fast that they nearly drowned before escaping. Everything was not going as planned for the Truro crew, it seemed like the island was just a bad luck place. There was one exception to this close tragedy that led to a very important discovery, the water that had entered shaft #3 was sea water! The only way sea water could have flowed into the shafts, which were hundreds of feet from the beach areas, had to be the work engineered by humans! There had to be a manmade tunnel leading from the waters of Mahone Bay to the area of the shafts! The flooding of the shafts was intentionally designed to prevent the discovery of what lay beneath in the pit! It was booby trapped!

The Truro crew searched for the source of the sea water intrusion along the shoreline at what is called Smith's Cove on the south end of the island. Where they discovered evidence of water movement during the changing of the tides. Water would seem to seep out of the beach soil at low tide, like it was draining, then would do the opposite, seeming to absorb water at high tide. What was done next in an effort to prevent water from reaching the beach area was the building of what is known as a cofferdam. A cofferdam is a type of walled construction that dams the perimeter area of the beach needed to block the water from rising into the seepage area inshore. The cofferdam was built in the shape of an arc that was placed away from the beach out to the area of the low tideline. The dam would encompass enough of an area to block the tide from rising back into the area where the water seemed to infiltrate the beach soil. The dam required a lot of work, but it was worth a try as there needed to be a way to stop the water from reaching the

Money Pit. What happened once the beach was no longer being flooded by the tides allowed the crew to begin digging around the beach area where they discovered what are known as box drains. A box drain is a system that is placed in the ground that permits water to flow into a pipe, tunnel, or holding basin as the water levels rise or fall. A similar system in use today is like the storm drains that are found alongside local streets. Upon further search of the beach, the crew discovered that there were five such boxes with each one channeled in from the bay waters into one main tunnel that seemed to lead landward toward the area of the Money Pit. To uncover the entire system would probably require a lot more manpower plus extra digging equipment that the Truro Company was not prepared to invest. They believed what was found was the answer of how to stop the water from flooding into the shafts.

More bad luck would strike the island again as a terrible storm had caused sea swells in Mahone Bay destroying the cofferdam that had once protected the beach search area. The storm would also hide the box drains under the surface of new beach soils that had been washed ashore. The island had prevented success once more for anyone who tried to uncover its mystery!

However, the island did reveal the magnitude of the challenge required to succeed. The discovery of the box drains was proof that they were intentionally created by a highly intelligent, highly skilled labor force. What must have taken not only days or weeks to engineer, but probably months of needed effort were likely required for the work of a small army to accomplish. The small Truro Company would be no match! Funding for the venture had been used up with the expense of a costly cofferdam that had now been destroyed. There were no new investors willing to risk their capital on chasing a failing dream that teased all with the promise of getting rich quickly. The Truro Company was broke calling it quits in 1854. It seemed that the many unsuccessful efforts to discover the treasure of the island had resulted in one unanimous consensus among the group, the island was cursed!

It would be almost another decade before another group would try to unfold the mystery of how to retrieve the treasure of Oak Island!

Tragedy Strikes!

The treasure quest that had taken place on Oak Island was yet to yield what the investors along with the trio of McGinnis, Vaughn, and Smith were hoping to retrieve. They all had hopes of finding piles of gold doubloons, heavy silver ingots, Spanish pieces of "8," or maybe some precious sparkling gems. So far though, the searchers had not produced what had been hoped for, costing much more money to fund their dream then they had anticipated. The loss of precious time away from their families, the long hours of tiring laborious work all of which had led some to suffer a broken human spirit. The many failures and setbacks were far too many. The dreams of what might have been found were shattered by the constant flooding of the shafts filling with water which often led to frequent cave-ins of mud and debris. Everyday when the crews would enter the shafts, there was always the sense of risk when work would begin, the next mishap could be the one that ended in a fatality. The island had to be cursed. There were too many things going wrong that should have been manageable. The good days were but few.

Not all had lost hope though in the search to discover what was now known as the Money Pit on Oak Island. There always seemed to be another adventurous group ready to give it a try. The stories had long been spread throughout the Nova Scotia Peninsula there was a search for a treasure of an unimaginable wealth that was just waiting to be found. Rumors of what the man named Pitblado had hid in his hand that day, plus the story about someone finding a gold watch chain during another dig on the island, were surely getting the attention of many curious listeners. The frequent talk that someone in the area had once actually used Spanish gold coins to purchase necessities had led many to become believers about the possibility there actually was treasure on the island. Treasures had been found in other places during this time era. So why not on Oak Island?

The Oak Island Association would be the next group of adventure seekers who would begin to organize. This group would obtain financing for their venture by offering shares in the company at $20 apiece with about one hundred willing investors accepting the invitation to join. A few of the former Truro Company people were still involved who were not ready to give up on the chance for their dreams to not come true. So the new challenge began in the spring of 1861 and would be the largest gathering of manpower and equipment to date.

The plan this time was the same as had been tried in the past, they would proceed to excavate another shaft near the Money Pit, approaching what was thought to be the treasure location from below in hopes to avoid the flooding water. However, the group was limited to what was available in those days, and the best way to challenge the flooding was to use a horse-powered pump that was capable of removing water from the depths of the shafts. The horse-powered pump would give the group a mechanical advantage in productivity, more so than trying to rely only on manpower. How the pump worked was by physically walking hitched horses around a geared wheel, which in turn would rotate a shaft connected to a smaller wheel that was wrapped with a rope. The turning wheel would move the rope up and down with water buckets attached to the rope at intervals. Using the horses to power the pump in moving the heavy loads of water up from the depths of the shafts, allowed the men to do any necessary digging. The rigs were used to bail out the money pit plus the new shaft. It would still take days of working in shifts around the clock to lower the water level in the two shafts. The digging crew that consisted of many men had been working on the new #6 shaft, when at about 118 feet down, they began tunneling toward the Money Pit. At that depth, the new shaft walls began to leak water. The workers had to quickly evacuate that shaft for fear of losing their lives. With the work now stopped due to the flooding of the new shaft, it wasn't too long afterward that a very loud crashing sound like that of falling wooden planks used for the scaffolding was heard from inside the Money Pit. Rushing over to see what had happened the crew discovered that the Money Pit had completely collapsed inward, filling with water, mud, and thousands of feet of lumber that once was used to shore up the shaft walls. The Money Pit had suffered a complete implosion! This was a catastrophe! The cause of the implosion was a factor of several things happening all at once. The digging that was being performed in the different shafts near the Money Pit, plus the tunneling that had been directed under the Money Pit along with the constant in-flow of water into the shafts, had made the ground dangerously unstable. One can imagine that a cloud of defeat surely hung heavily over the island once again as all had witnessed the destruction of their labor.

It was nearing the end of summer in 1861 when this had happened. There was now a need for more money to pay for the much needed work to clear the area of the Money Pit and the clearing of the other caved in areas. The determined investors did come up with more money while planning another approach to conquer the flooding problem. This time they would use steam powered water pumps that were more efficient in removing water from the shafts. By the time the pumps were readied on the island it would be the fall of 1861.

Boilers used in olden days were not well regulated as they are by today's manufacturing safety standards. Quality and construction standards had no stringent guidelines to insure that the boilers would be safe. What happened to one of the boilers being used is that it failed under the high pressure of the steam buildup causing it to unexpectedly burst. That eruption of scalding hot steam and water engulfed a worker who would die from this horrible incident. Several others who were nearby were also injured prompting the decision to shut down all operations. Between the Money Pit cave-in, the other shafts constantly flooding, and the first death on the island, it all had fostered into a good reason to quit! Working on Oak Island had become too dangerous. Little did they know of the curse by those whose bones would later be found at the bottom of the Money Pit!

This group of investors that had formed the Oak Island Association were determined to not quit though returning in the spring of 1862 to the island with renewed financing that would permit the hiring of a new crew. For the next two years, efforts were made to reach the Money Pit from below by repeating the same techniques of pumping water out of newly dug shafts around the area of the once known location of the Money Pit. Unfortunately the results were never positive. There would be no success!

The Oak Island Association would go broke, finally calling it quits in 1864. This would be one more venture defeated by an island plagued with mishaps, failures, and now death. They would abandon the island, all the equipment was now gone, the commotion of human activity had ceased. An eerie quiet would soon consume the island. It was now only the scene of remnants left behind from nine unsuccessful shafts in the area that once was thought to be the treasure laden Money Pit.

The year 1866 brings yet another group to try their skills at uncovering the supposedly hidden treasure of unknown fortune that one can only speculate when it was buried or by whom. This time the venture group is called the Oak Island Eldorado Company, another group financed by investors with each purchasing shares costing $20 apiece for a total of 200 shares. The group would also be known as the Halifax Company getting that name from where they had formed in Halifax. The goal of this group was to prevent water from flowing into the Money Pit before attempting any more excavation efforts which seem to fail as soon as they would near the Money Pit area. The idea was to build what is called a coffer dam around part of what is known as Smith's Cove located at the end of the island, about 500 feet from the Money Pit. That was where the Truro Company had found the tidal inflow and outflow of water on the beach concluding that it was the source of the tunnel flooding. This would be another attempt at a cofferdam which the Truro Company had tried in 1850. The dam type structure would be erected off shore to block the beach area from any tidal inflow. It was a good idea to try to block the water from flowing into the flood tunnels before any work on the pit would begin. The effort to build the cofferdam was no easy task since it had to be constructed using only materials that were readily accessible in the near area which were rocks, clay, and island lumber cut by hand. The cofferdam would be about 375 feet in length by some 12 feet in height in an arc shape about 110 feet from shore. Once the cofferdam was constructed, the water inside the arc area would be pumped out thus preventing any sea water from entering the tunnel shafts. A labor intensive undertaking to say the least, but it was a very logical approach to solve the problem of the persistent flooding.

Unfortunately after all the work had been completed, stormy weather would come to the area that created plenty of rough tidal waves, eventually battering the cofferdam badly enough to cause it to collapse. To the dismay of the investors, all of the expensive costs for that construction was for naught.

The Halifax Company was still invested in the treasure hunt with no intention of calling it quits after the failure of the cofferdam. So once again there would be one more attempt at creating a new shaft near the Money Pit shaft, again trying to tunnel laterally under the Money Pit in hopes to avoid the flood tunnel. But, as before, they would be defeated by the flooding of the water rising into the pit. By 1867 the company had exhausted its' funds deciding to call it quits for good. The only positive result that the Halifax Company achieved was knowing that the flood tunnel system leading from the beach to the Money Pit could be controlled by the use of a cofferdam construction. Too bad it had failed!

This time, there would not be another organized search on the Island until 26 more years would elapse. It was said during this time there was some silver Spanish coinage used to purchase things acquired by Anthony Vaughn, one of the original trio who was living on the island. Could it possibly be that Vaughn may have been doing some searching on his own, finding a treasure site on the island other then that in the pit area? Yet one more mystery open for discussion.

The Curse of Death Again!

There had not been any organized attempts since 1867 at trying to conquer the Money Pit. Maybe the knowledge of the failures by those who had previously tried was enough to keep anyone with a sane mind away, who did not want to waste their money nor time. But there always seemed to be a new adventurer or two willing to ignore the history of failure by others. When attempts by others prove fruitless, there are others who see their opportunity to succeed. So once more, another adventurous group would be formed by a few who had knowledge of those earlier attempts. A new name in 1893 would now be heard by the curious others: the Oak Island Treasure Company.

This new group would be joined by two individuals who would eventually become the most involved for the longest time yet. The first name to mention is Frederick Blair, the nephew of Issac Blair, a man who had worked on the island during previous attempts. Blair would bring with him an engineer named Adams Tupper who had designed a new plan of attack to avoid the flooding problem that was gained from knowledge of the previous failures and successes. Another person that will become deeply involved for many years to come is William Chappell, a lumber businessman. A couple more investors would join the group in purchasing newer equipment than what had been used in previous attempts to conquer the pit. One piece of the newer equipment was again another steam powered water pump for siphoning the flood water from the pit. This group would try for almost two years to defeat the flooding of the pit while continuing their digging efforts of trying to use an adjacent shaft to gain entry into the money pit. It just seemed like the island would not co-operate with the group attempting to get the pit dry enough to safely gain access. The water would continually seep into the pit requiring a never-ending struggle to get it dry. Unfortunately, there would be another worker who would lose his life when he fell into the flooded pit while trying to ascend a rope. This drowning in 1897 was now the second death to have occurred by a worker on the island, prompting the current workers to call it quits with their belief that the island was truly cursed as being a death trap.

Work however would resume after the hiring of a new crew that would lead them to the discovery of some significant new clues. These renewed attempts would give a sense of belief there really was something buried in the depths of the pit. Using the technique of drilling exploratory shafts down into the area of what was thought to be the pit had revealed there was some type of a hard obstruction encountered at a depth of 120 feet. What the drillings also discovered was this type of obstruction contained pieces of wood and what appeared to be a substance similar to that of cement which had been found on the ends of the drill bits. The obstruction was thought to be several inches thick like some kind of a walled structure or box-type vault. After drilling through the wall, they hoped to enter into a void. This surely was believed to be some type of a manmade object with the purpose to contain something of value, like a treasure. I am certain the crews believed what they had found was the legendary treasure once hidden by pirates. While examining the substances recovered from one of the drillings, a small piece of parchment with a single, squiggly letter like "V" written in ink was also discovered. Could this be a piece torn from some ancient writing that would tell of the treasure below? Perhaps! That parchment is yet to be identified. Yet another mystery left to be solved!

The flooding of the pit and the nearby shafts would still be troublesome. There had to be an answer to stopping the constant inflow of seawater in order to safely dig downward. In an attempt to locate the source of the flooding water, a red dye was introduced into the water of the pit. The red dye color would spread throughout all of the water found in the pit and into any water sources connecting to the pit, hopefully to appear at any intrusion location with the changing of the high and low tides. What this process unveiled was the evidence of red dye flowing out onto the beach area at the southern end of the island. That location is what is known as the Smith's Cove beach area. This proved there was definitely seawater intrusion as was believed back in the 1850s when the Truro company first discovered there was water seeping out from under the beach. The beach wetness was greater than normally evidenced by a natural flow of drainage through heavy sand and soil types that are found between the beach area and the pit. Those soil types would almost certainly block that amount of water flow. There had to be some type of a manmade conduit or tunnel that could allow enough inflow and outflow of water to flood the shafts with the changing tides and then to flow back out to the beach area. Something had been intentionally designed to impede anyone from getting to the bottom of the pit! The evidence was clear. There was no doubting that there was in place a very cleverly engineered system created in the past that was deliberately designed to stop anyone from being successful at finding what was there.

It was now 1899. The Oak Island Treasure Company was no closer to success than when they had first begun. There was proof though that the pit was deeper than thought earlier, maybe even as deep as 160 feet. Their assumption would be quite right as the future would reveal. There were often starts and stops because of weather, along with the crews occasional halting of the work on the island to pursue their seasonal livelihoods of farming and fishing. This inconsistent work activity had left those involved thinking their efforts were not really serious enough. The work by the group was further hindered by the constant failures to make progress in the pit because of the frequent cave-ins of the clay walls. These delays and setbacks were only frustrated further by the unabated flooding of the shafts. The failed tunneling attempts to undercut the pit had signaled a big red flag. By 1900, the island would lose all of its current investors, except for one, Frederick Blair. Without getting into all of the missteps yet to come, the island would continually be troubled by shaft cave-ins and insufficient funding. Blair held on as best he could, attempting to continue the work with occasional help that would be financed by small groups of investors with limited capital. Years would go by, even the event of World War I would pass, but finally in 1931, there would be new hope in the arrival of a former investor and friend.

William Chappell had been here before as an investor/searcher who knew of all the failures, yet he was back to give it another try. The pit shaft would again be cleared of debris as far as possible to a new depth of about 160 feet. Some sample search drillings were then made at deeper depths where they discovered that there was some type of an obstruction down deeper. This obstruction was thought to be something like a large vault or a solid walled container like box. Could this be where the treasure laid awaiting its retrieval? Maybe. This discovery will be referred to later as the Chappell Vault.

In Nova Scotia your ability to do work in outside situations are definitely limited by the weather of the north. Intending to return in the spring of the following year, the Chappells and the Blairs believed their next attempts to find the treasure were close at hand, but as things go on Oak Island, it would not be so.

The owner of the island land where Blair was searching had died in 1931 and the new owners/heirs were reluctant to give him a renewal of his lease thus causing a rift between the parties. What some would call extortion, the new owners wanted Blair to buy the land at what was considered an overvalued price during those times in order to have access to the area. He would refuse, thus stopping any of his further pursuits. Chappell couldn't help either at this time as he had invested all of his monies with no return and did not want to jeopardize his lumber business on the mainland. Now there would be yet another lull on the island.

As usual, time passes and there would come another adventurous sole seeking to pursue the dream of finding what others claimed to be just within their grasp. Oak Island seemed to be like the old teasing trick of dangling a carrot in front of the horse to tempt him into moving, but this carrot was the lure of the island's treasure to entice the searcher. This new challenger was a man with deeper pockets who could afford the risk. Having been successful as an entrepreneur in the steel fabrication industry, a stint in the insurance business, and later as an automobile dealer, this was the type of adventure that would fit his resume.

Gilbert Hedden was an educated man who had attended some specialized schools in applied sciences and technical arts. He had the business savvy along with the confidence of prior success to attempt what others had abandoned. Hedden knew of the history about Oak Island and would break the lull in activity by buying lot #18, the land of the pit area that the heirs possessed. Hedden would continue to have Blair and Chappell involved as their knowledge of previous operations remained very valuable, plus Blair still held a valid treasure searcher's permit that had been issued by the government. Hedden would bring in a new group to remove the water and debris from the current shafts in efforts to dig a new shaft next to the Chappell shaft. After doing the repairs on the Chappell shaft, now at a depth of 170 feet, the new larger Hedden shaft was dug next to it stopping at a depth of about 124 feet. At that depth, they decided to drill test holes downward to examine what they would encounter. The Oak Island curse would come haunting once again as money became an issue with Hedden, who was suddenly faced with big money decisions in 1938 about his auto dealership along with him owning back taxes. Hedden would cease his direct involvement on the island, eventually going almost broke. Hedden, who had high ambitions to succeed would become just one more defeated Oak Island searcher. Hedden would leave the searching to the duo of Blair and Chappell. He would still hold on to his dream hoping someday to return to his quest, but he would finally give up and sell his land on the island to another investor in 1950 for just $6000. Ironically, all of the land that Hedden had owned including eight other lots eventually will be owned by the Chappell family. That land is the area where the current Money Pit is located and now could be worth well, you can only imagine. I guess only time will reveal its true worth.

The biggest gain for Hedden in his exploration of the island was the exposure of the stone triangle that was found in 1897 by Captain Welling.

So Hedden had become a sidelined searcher out of the picture, but is quickly replaced in 1938 by another dreamer named Edwin Hamilton, an engineering professor from New York. Hamilton, Hedden, and Blair would enter into an agreement of partnership with Hamilton as the leader still using the same group that Hedden had hired for the clearing of the pit and any drilling efforts. The engineering background of Hamilton was put to the test as he directed the drilling group to try to locate the once known position of the Money Pit. The exact location of the pit had become uncertain after the 1861 shaft collapse that filled the area with tons of dirt and tunneling construction debris. The drillings would be directed at angles that could explore more of an area instead of drilling vertically downward in hopes to intersect laterally any previous tunneling.

For the next five years, drilling would continue along with the deepening of the Chappell shaft to a depth of 200 feet. The same old results that others had encountered were still evident during these searches, discovering only more pieces of oak wood from previous manmade shafts and tunnels. In 1941, the second World War had begun causing many to change their plans, and in 1943 Hamilton would end his adventure on Oak Island to enter into a more promising venture in the boat building business.

There was always an undying interest in Oak Island by anyone who had heard of the proclaimed possibility there was a pirates valuable buried treasure awaiting to be found. After the war there were things like lives to rebuild and time to start new ones. So once again the island for awhile would be quiet from the noise of men poking the earth with shovel and machinery.

Mel Chappell, the son of William, would enter the picture at just the right time at the beginning of the 1950s. Chappell would join forces with Blair as both men had knowledge of the ups and downs that had become the norm for any Oak Island venture. The younger Chappell had worked on the island with his father back in the 1930s. Melbourne R. Chappell was a person that seemed to have good fortune on his side as he was able to acquire the land in the Money Pit area at a reasonable price from the person that had bought it from Hedden. Some people get nervous about keeping uncertain investments too long prompting decisions to sometimes sell their holdings to escape a loss they can't afford. While Chappell was fortunate to acquire such a vital piece of the island, Blair was not able to fully witness what Chappell would accomplish as Blair would die at 83 years old in 1951. Mel Chappell now was the only major treasure hunter with a valid government search permit. For a few more years Mel Chappell would tolerate wannabe investors, dreamers with theories, peculiar individuals, and just plain nosey people. When you have something like a mystery to solve or a treasure to find, there is always someone who wants to get in on the action. Then there were some, thinking they knew how to do it better than Mel, or those just wasting his time with gab. What Chappell needed was another serious investor with the funds to back up the talk.

Enter the oil man from Texas with the big dollars who was associated with a group of other well funded investors. George Greene was a cigar smoking man who wore an expensive Stetson hat and typical cowboy boots, all the image of a successful Texas oil man. It was now 1955, things in the country had improved after the Korean war that began in 1950 lasting three years. There was even better machinery now that could be bought to do the work needed on the island. Mel Chappell was probably elated that he now had the opportunity to further his exploration in locating the elusive Money Pit. The much needed aid from some newfound moneyed investors was what he had hoped for. The duo would begin their work by first drilling new bore holes into the area thought to be the Money Pit, only to encounter the remnants of wood that was used in the various shaft walls and work platforms left from past failed attempts. Greene had made a plan to locate the so called vault of the money pit that would identify where they needed to excavate. What the drillings did find besides all of the previous debris from earlier attempts was the evidence of void areas at depths below 112 feet plus other open spaces that had bottoms down as deep as 180 feet. No one was certain what all of this meant. Were these spaces created by man or nature? They weren't sure what it was they had discovered. The attempts of the oil man to find the obstruction thought to be the vault had failed. It was now the fall of the year, and after only spending two months on the island, Greene would leave planning to return in the spring. A new oil deal opportunity would arise though, permanently changing Greene's future intentions to return to the island.

It was 1958, the island seemed to have a never-ending supply of wannabe fortune seekers awaiting their turn to solve the mystery of the Oak Island treasure. Two brothers from Canada who had visions of great wealth buried on the island would be next to accept the challenge. They had imagined like many others that the treasure could be in the hundreds of millions of dollars. As with any legend about treasure, the jackpot seemed to continually grow over the years, making the chance of failure seem but a minute risk to gain the imagined reward. The Harman brothers would do what others had done before them in drilling around the suspected area of the Money Pit in attempts to locate the original site. Their efforts proved no more successful than any others had and they would leave empty-handed after just a few months of searching. Their dreams of riches would fade almost as fast as their funds.

Entering the challenge this time would be one of the most aggressive searchers to date who would lose the most in the history of Oak Island! It was 1959 when the family of Robert Restall would begin to move to the island and start the search of the Money Pit that would eventually end in a family tragedy. The Restalls were a married duo of performing motorcycle stunt riders that had traveled across America and Europe for about 20 years before deciding to take on the challenge of their next risky adventure – Oak Island. The act that the married couple performed was to ride their motorcycles inside a large 20- to 36-foot diameter open top barrel like cage at high speeds to entertain audiences around the world. The Wall of Death, as it was first referred, was usually constructed with wooden walls that could be dismantled for travel to be rejoined at the next venue. Observers could watch the act in the barrel from a catwalk around the top of the circular cage. Back in the 1950s that was quite a daredevil feat that was usually featured as the main attraction at traveling Carnivals and Fairs. Robert was now in his 50s with dreams of doing something not so tortuous to his body or lifestyle. The family would give up the traveling life after Robert took a job as a regular working man in the steel industry in Ontario for a few years. Robert had known about Oak Island years before when he attempted to enter into an agreement with Chappell to conduct his own search.

Unfortunately for Restall, others had been in control at the time making him wait for their departure. Restall had heard of the Harmon brothers leaving and he quickly approached Chappell to get his turn at the treasure. His funds were limited to what savings the family had, yet his goal was to do what he could do even though he was unable to buy or rent any necessary heavy machinery. The first year or two, he and his eldest son, Robert, now 18, would use picks and shovels to try to excavate the previous shafts created by Hedden and Chappell. They would live in very meek conditions sheltered only by a couple of small shacks without electricity, water or other plumbing facilities. Their efforts would however be aided later with some financial support from some old friends enabling Restall to buy a much needed water pump that Hedden once owned. Work was now a little better when water could be removed from the shafts permitting the Restalls to re-wall much of the two shafts as possible. But Robert still faced the nagging problem of water seepage into the shafts and had begun to dig a pit between the shore and the Money Pit in hopes to divert the water away from the area.

It was the summer of 1965 that found the Restall team digging a shaft from the beach area of what is known as Smith's Cove toward the area of the Money Pit. Robert had been using a gasoline pump on the this shaft of just 27 feet deep when his son, Robert Jr., discovered his father lying in the water at the bottom, seemingly unconscious!

Robert Junior climbed down to aid his dad when for some reason he also became unconscious. It didn't take long before a teenage worker and another partner named Karl Graeser would climb down into the shaft in an attempt to rescue the two. At the end of the day, four people would die at the bottom of that shaft thought to be filled with poisonous carbon monoxide fumes that had apparently drifted down from a gasoline powered pump. A tragedy that should not have happened. Could it be the supposed curse of the island bringing death once more to those seeking its treasure? Today, there is a memorial on the island with the names of those that died that day. They are: Robert Restall 60, Robert Restall Jr. 24, Karl Graeser 38, and Cyril Hiltz, 16. The island had now claimed a total of six individuals in the search for the buried treasure which in itself remains a mystery as to what it might be.

Robert Restall would leave behind his wife, Mildred, and their youngest son, Rick, with an uncertain future. Was the island really cursed?

A Big Dig

In the next few decades to come, Oak Island will become one of the most intriguing mysteries about the search for buried treasure in modern times. It is engulfed by not just one question, but by many, that create something more of a puzzle needing each piece to be answered to fit with the next to complete the mystery. This thing about Oak Island could be more complex to finding a solution then the ones about the disappearance of Amelia Earhart, why was J.F.K. really shot, where is Jimmy Hoffa the mob boss, and where or what is the Holy Grail? There are more mysteries besides these yet to be unraveled, but after the printing of the Oak Island story in the publication of the *Reader's Digest* in January 1965 written by David MacDonald, the island's mystery would begin to be elevated to a new place among mysteries. Although the book was distributed before the tragedy of the Restall family in the summer of 1965, their deaths surely heightened the public awareness that maybe there really was a buried treasure that was cursed and not meant to be recovered. Many had already tried to retrieve what had been secreted deep on this strange shaped island that was so craftily protected by the booby trap flood tunnels. Now the world had become more aware of a potential great fortune that was only awaiting to be claimed by those adventurous enough to risk it all. The story was out for all to know.

The Restall accident wasn't going to be a deterrent for the next individual in attempting to implement his plan of attack against the difficulties evident on the island. It was October of 1965, just a couple of months after the Restall fatal accident, that Robert Dunfield would step foot onto Oak Island. This would be another oil man from California with expertise as a geologist. He would approach the treasure search of Oak Island in a new and very damaging way as some would argue. Dunfield would use a different type of equipment to do his searching. He brought in bulldozers and big earth digging cranes. To get the big machinery onto the island would require the construction of a causeway to be built from the shores of the rural community of Western Shores to the island. The only way to the island before 1965 had been by the use of boats or barges to ferry equipment and crews across the waters of Mahone Bay. The time had come to build a permanent access that could be used for the transport of that bigger machinery and to allow easier and safer travel to the island, especially during periods of inclement weather. The causeway (type of raised road) of stone and fill dirt would be built to cross approximately 600 feet of open water from the mainland at Crandall's Point to the island's shore access located at its northwest end. The causeway was quickly completed with the blessings of Mel Chappell, the owner of the designated landing site on the island.

Dunfield also had some other investors backing him. One of whom will become very well known in the Oak Island story, Mr. Dan Blankenship, a businessman from Florida who would eventually move his family to land he would own on the island. Dunfield's plan was to use his bulldozers to initially clear an area several feet in depth around the location of what was thought to be the Money Pit rather than to excavate all of the debris that had accumulated over the years from any previous shaft cave-ins. The plan was with the hope of clearing away all remains that would be misleading to the actual location of the money pit, which by now was uncertain at best. This plan was successful when the operation did uncover some of the original wood cribbing that had been used for the walls of the Money Pit. However, as Oak Island does, the usual flooding of the pit would soon require the need for a massive water pumping rig to siphon out the water while excavating the pit. The scene around the pit area was beginning to resemble one of a sizeable borrow pit operation with the big bulldozers and a giant excavating crane digging down and moving tons of earth away from the site. The area was shaping up to be more like a very wide open pit that eventually would measure about 100 feet across and almost 150 foot deep! This had become a massive assault on the landscape of the island that was producing nothing more than the retrieval of old wooden timber debris from all of the earlier constructions used in shaft wall cribbing and damaged metal casings once used to guide drilling bits. Winter had now come, and the weather of Nova Scotia had dropped so much rain that the disturbed area around the dig had become so unstable that a cave-in of all the pit walls occurred.

Dunfield would retry in the new year to find the original Money Pit by once again using his bulldozers and big cranes to excavate around the area thought to be the pit. There would be some successful drilling used as well this time that would find a large 40 foot void at a depth of about 140 feet near the big dig. This could have been the same void found in the earlier probes that some believed was possibly a manmade chamber or vault. There was still the worrisome problem of water seepage into the pit area that had to be blocked to enable further searching. Dunfield was yet to have the success he needed prompting him to become more aggressive to find a solution to the flooding problem. He would now begin a search of the island to discover where the water flow could be blocked. The operation would first dig along the south shore in attempting to unearth any type of evidence indicating a flood system, finding nothing that was helpful. The digging would however destroy the stone triangle site that was found by Captain Welling in 1897. The operation would then try the Smith's Cove area only to destroy any earlier discovery of the box (finger) drains on the beach that were discovered by the Truro group in the 1850s. Nothing had worked to stop the flooding.

Dunfield had given the island what he thought was his best effort at finding the original Money Pit. What Dunfield had really done was to destroy valuable evidence that held any historical significance vital to understanding what had happened on Oak Island. Thankfully there is enough recordation by past searchers that had witnessed those historical facts. The island had been ravaged by what had been done with the use of those big bulldozers and giant earth digging machines. The landscape had been dramatically changed, some historical artifacts were damaged or destroyed, and the exact location of the Money Pit was still unknown! Dunfield was done; the investment was a bust. He had joined the group of the Oak Island humbled and he returned to California to stay.

Dan Blankenship would hold on to the lands acquired in the partnership while also retaining any agreement with Mel Chappell.

A New Era Begins

The year 1965 marked a milestone in the history of Oak Island in that its mysterious story would be well exposed after the release of a couple of highly publicized writings. The *Reader's Digest* article published in 1965 titled "Oak Island's Mysterious Money Pit" written by David MacDonald was printed in a January edition that had a circulation output in the U.S. of over one million copies. Another earlier publication in book form was written by R.V. Harris titled *The Oak Island Mystery* that came out in 1958 with reprints starting in 1967. These two publications along with some shorter news releases would prompt the involvement of some of the most recognizable and important people who would come to be the most enduring searchers yet.

The publications about Oak Island had now revealed to the readers a story fraught with mystique about something that had been secretly buried hundreds of years ago that no one knew what it really was, nor whom had done what, nor when. To some this would be the beckoning call for them to unravel the mystery that had already been attempted by others in the past. The challenge to the next curious souls in line was to succeed where others had not! The next adventurous ones would try to solve all the questions, and to hopefully find a treasure more valuable than a king's ransom or the Kingdom. This would be their turn!

Mel Chappell was already an important player in this mystery to be solved, but he was about to be joined by another important player who had been waiting eagerly on the sideline to enter the story. Fred Nolan was a land surveyor by trade who lived to the north of Oak Island in Halifax, Nova Scotia. Nolan had heard of Oak Island from the locals in his travels while doing his surveying work, but his interest had been peaked, as has been told, after reading the R.V. Harris book. Nolan would visit Oak Island to speak with Mel Chappell about wanting to do some searches of the island, but Chappell at first had been hesitant to let anyone start snooping around as Robert Restall had still been involved in his search at the time. The island had already been surveyed when the land was sub-divided into lots in 1762 and partly again in 1937 by Charles Roper. Chappell did eventually let Nolan do a new survey of the land, especially since Nolan had volunteered to do it for free.

So begins the examination of the island by Fred Nolan in the early 1960s. What his discoveries will unveil will become some of the most historically important to date. I will temporarily skip any further detailed narrative here about Fred Nolan as later there will be a in-depth chapter about what his most important role is pertaining to this book. What his notable find uncovers needs to be highlighted on its own. What will be done here is to briefly describe his relationship with the other participants on the island as time progresses.

Fred Nolan would invest a lot of time and money into his free surveying work of the island, so much so that his business would sometimes suffer from his absence. He was a determined man looking for the answer to the mystery of what the island was hiding. In his desire to unravel the secret, he wanted Chappell to give him more involvement beyond doing the free survey work, even requesting permission to do some digging on his own. The activities of Restall and Dunfield would temporarily delay any searching by Nolan who by this time had aggravated Chappell with his persistent requests. Nolan though was a determined man in his mid-30s with a sharp mind who decided to check for himself if there was any way to get access to the island without the permission of Chappell, the supposed owner of the tract of land Nolan wanted to search. Nolan, who as a land surveyor would routinely examine land records for property ownership, decided to check Chappell's ownership on the island's land and discovered that the heirs of the original owners had not sold everything to Chappell. Nolan was determined to gain access to the island so he bought lot# 5 and lots #9 through #14 from the heirs. Nolan, now as a land owner, had the control of the easterly end of the island which encompassed the area to be known as the swamp. After Mel Chappell had been informed that Nolan had made the transaction it would become the cause of a very contentious relationship between the two men.

Chappell and his partners felt they had been duped by Nolan and decided to not let Fred Nolan on the island via the causeway that Dunfield had built. What Nolan had actually done was to outsmart Chappell and the others in doing what was perfectly legal. The backlash of his action would force Nolan to make his trips to the island from the mainland by boat in order to search his land. The feud was tense to say the least with Chappell even sitting guard on the beginning of the causeway at Crandall's Point with a firearm forbidding Nolan not to use it. Nolan, being the determined man he was, would purchase just enough land aside the narrow causeway at Crandall's Point to block Chappell's lease holder Dunfield from using it! Dunfield needed that access to move any heavy equipment back and forth to the island. Without going into all of the conflicts here, eventually there would be some wheeling and dealing between Nolan, the Triton group, a few lawyers, and a major Triton partner Dan Blankenship. A quasi-truce would be reached between the players allowing an amenable ingress and egress of the island by all parties. Fred Nolan had now overcome an impasse; he also was now permitted to safely cross to the island that he had hoped to explore. Explore it he would!

Dan Blankenship had become part of the Dunfield Triton group when he became financially invested in the operation in 1965. Eventually he will become the only remaining partner of the Triton group after Dunfield vacates the island. For the next few years, Blankenship would travel back and forth between his business and his home in Florida to Oak Island working on the island when weather and work permitted. He would eventually move the family to the island in 1975 where he would build a home on lot #23, a parcel he had purchased. Much of Dan's search operations were similar to what others had done before him, doing drillings, clearing old debris from the area of the Money Pit, and sinking new exploratory shafts. He even built a cofferdam around the beach at Smith's Cove in an attempt to discover any source of sea water flow to the Money Pit area. Dan did discover several items of importance in his years of searching on the island. One peculiar item found in his excavations was a pair of handmade iron scissors, not sure what they were used for though. There were other items like old, handmade iron nails, large wooden dowels, big wooden beams with roman numerals carved in them and even some coconut fibers found at the Smith's Cove beach area.

Probably the most significant thing that Dan discovered, or came to realize, was that the island was like a labyrinth of old tunnels under the island. He had done enough drilling and digging of the island to understand that the Money Pit was linked to other offshoot tunnels that had been dug much earlier than those left from the searches of Dunfield and Chappell. These connecting tunnels could be the infamous flood tunnels that continued to allow water to flow in from the finger drains at Smith's Cove.

Blankenship was doing a lot of work on the island that had cost plenty of money, so much so that he welcomed another investor in

1967. Dunfield was now gone and in his place along with a sizeable group of financial backers was David Tobias, owner of a successful business in Montreal. Tobias would buy out Mel Chappell's major land holdings in 1977 for $125,000. This relationship between Tobias and Dan would last almost 30 years with Dan doing the searching and managing of the project on Oak Island with Tobias delivering much of the operational monies. With the influx of new money, it didn't take long for the new alliance to create more search shafts around the area of the once known location of the Money Pit. These boring shafts were about six inches in diameter that would be drilled down to depths as deep as 235 feet. A major discovery was found at about 140 feet when the drill dropped into what was a void of several feet before continuing downward hitting bedrock at about 180 feet. The void or cavity could be manmade or natural, but it was suspicious either way. That bore hole was more interesting than any of the others to Dan Blankenship who would name it Borehole 10X.

Borehole 10X was puzzling to everyone: it not only dropped in one void, but would drop into two more before going down to 235 feet. A thin-walled, metal six-inch diameter tubelike casing would be inserted into the shaft hole to prevent any dirt along the walls from falling into the opening. A technique of using pressurized air to clear any debris after the insertion of the casing would then be used to force any debris to the surface. One day what was blown up from the bottom of the shaft were pieces of glass, seashells, some birds bones, and a hand full of some old thin type of metal that was similar to lead. The metal would be tested and analyzed dating it to be over 200 years old! That would mean the metal was processed from raw ore around the year of 1750. This would be a very important find that could possibly date the beginnings of the Money Pit!

Dan Blankenship was convinced they had found an integral part of the original Money Pit, or at least evidence indicating the possible location of the flood tunnels. In the years following the drilling of Borehole 10X, several others eventually would be done. There would be cameras lowered into the shaft all the way down to the bedrock found at the bottom of the 235' shaft. Dan even claimed that at one time while watching the camera monitor in 1971 that he had seen several objects similar to rounded top chests lying on the shaft bottom along with what appeared to be a human body. Was it really some treasure chests he saw and a human body?

Blankenship would enlarge the size of the six-inch Borehole 10X into a 27-inch diameter opening and insert a new metal casing shaft that was just large enough to allow a diver outfitted with an air hose to go down inside. Divers were hired to investigate the depths of 10X, descending past the new metal casing that ended at 180 feet, and then continue downward passing by the hard earth walls to the bottom at 235 feet. At the bottom the diver entered into the large void or chamber Blankenship had seen on camera, but after searching around in the murky water with almost no visibility, found nothing! There were no treasure chests, nor any human body as Dan once believed.

Dan would make several dives himself in the coming year, entering the narrow metal encased shaft wearing his diving gear. He would descend the encased walls down to 180 feet leaving the next 50 feet or so of the shaft uncased that had been drilled through hard natural minerals like gypsum and limestone. Dan's dives even allowed him access into the void found below the depth of the metal casing while his son, David, a grown man, would monitor the dives from above. Dan was a brave man who was now near 50-years-old still defying the fate of the curse that had already claimed the lives of six able-bodied men on the island. But Blankenship was determined to find what was below in the deep dark shaft that was filled with murky water. One such dive would come too close to claiming Dan Blankenship's life as being the seventh person to die in the quest to find treasure on the cursed island. Dan was down about 145 feet in shaft #10 on a dive one day when a strange loud metallic noise alerted him that the metal casing of the shaft walls were beginning to buckle inward. Probably from the walls of the casing not being thick enough to resist the pressure from the surrounding earth. Fearing this was going to be a collapse of the entire shaft around him if he didn't get out in time, panic set in and he began to yell commands to David to haul him upward as quickly as possible. Dan Blankenship luckily avoided his demise as David urgently hauled his dad upwards out of the shaft just as it imploded beneath him.

Dan Blankenship had escaped a death trap that day, and the experience would not be forgotten. Returning the next day he and his son David found that the shaft had mostly filled in with debris leaving only 65 feet from the top clear enough to enter. The shaft that once was accessable down to 235 feet would now have to be re-cleared once more. Dan would have been buried alive that day had his son David not been present to rescue him. The legendary curse of "seven must die before the treasure is found" did not come true that day, but it would continue to haunt the minds of men, especially Dan.

Dan would not attempt the clearing of shaft 10X again until the year 1978. Dan wanted to do one more big attempt though, this time using larger eight-foot diameter cylindrical steel tank cars used by the railroads. The bottoms would be cut off the tank cars changing them into giant steel cylinders. A clever idea verses having special cylinders built in a factory. This is only one example of Dan's ingenuity used in his many years of search efforts displaying the type of knowledge needed to tackle the complexity of the island's many challenges. The large steel tank cars would be welded together end to end and positioned into the re-dug shaft down to ninety feet. The steel tanks would create a new stronger, wider wall permitting safer dives to the bottom of 10X. There would be some dives down into the newly enlarged shaft but all produced nothing that was hoped. Financial concerns over operational costs had always thwarted progress on the island, and once again it would be just one more thing to cause the renewed efforts at 10X to be halted.

In 1983, a property rights argument between Fred Nolan, Dan Blankenship and Tobias would be tested in court causing all parties to cease any of their search work on the island. Years would pass by while awaiting to renew work on Borehole 10X, but hopes would be abandoned in 1993 until a renewed effort would hopefully arise in the future. Money shortages, legal delays with the government, disputes with other landowners, equipment failures, weather set backs, and the continuous problems with shaft flooding would seem to defeat anyone, except Dan Blankenship. Blankenship had spent decades in searching the island in the hope that 10X was the answer to the discovery of the phantom treasure. He had invested too much to quit, so he would wait.

Years would pass by again before Dan Blankenship would see the type of new investors he had hoped for that would help to continue the story. They came this time arriving from the state of Michigan. This new group would be the brothers Marty and Rick Lagina along with their partners Craig Tester and Alan Kostrzewa. The Lagina brothers were both young boys in January 1965 when Rick at the age of 13 would read to his 10-year-old brother the story about Oak Island that was published in the monthly magazine, the *Reader's Digest*. From that day forward it had become a fantasy for Rick that someday he would go to Oak Island to see what the mystery was all about. That fantasy has become a reality today.

Rick's quest started to come true in the form of an opportune event when David Tobias decided to sell a piece of his Oak Island land holdings in 2005. Brother Marty had been a very successful business owner in the field of natural gas and energy who had recently sold his company receiving a substantial buyout. The brothers decided to purchase the piece of land Tobias was selling thus establishing their presence on the island. Yes, Tobias was a partner with Dan Blankenship and one would think Blankenship would have been first taker of the land deal, but not all seemingly obvious things are actually so. Tobias and Dan Blankenship had been partners for decades, but over time had developed disagreements about operational objectives, mostly concerning Dan's obsession with Borehole 10X. Dan did have input in the sale though that would ultimately favor the Lagina brothers. The Lagina brothers and their partners would eventually buy all of the Tobias holdings, along with acquiring part ownership in Tritons Treasure Trove License giving them a partnership with Dan Blankenship. Today the Lagina brothers own 26 of the island's 32 lots, leaving the Blankenship family and the Nolan family the remaining six lots.

As time allowed, the brothers would eventually build a working relationship with Blankenship that would enable the combining of resources to continue the challenge of solving the mystery of the island. This new alliance between Blankenship and the Michigan partners would unite under the corporation title Oak Island Tours that Blankenship and Tobias had formed earlier. That venture was designed to accommodate the tourist interest in the island while providing income from tours and visits to the island's small history museum.

The first actual joint searches between the brothers and Blankenship didn't begin on the island until 2012 because of legal delays concerning government permits pertaining to treasure searches and entitlements. Finally with permits in hand, the Lagina brothers and others would begin to renew the searches of the island bringing with them new technologies and enthusiasm. Rick Lagina was surely elated to finally get things going as his fantasy was starting to become true.

The first couple of years were slow to realize any real progress that was any more productive than what had already been accomplished on the island. More drill hole searches would be done trying to find the whereabouts of the Money Pit while the Lagina team would learn more about what the previous operations had discovered. It was more of the same repetitive efforts that seemed to be the only way to proceed.

Oak Island had always been a mystery that would occasionally capture a news line somewhere or be the topic of a new book release written by another captive author. It had been sometime since any real worldwide exposure had been captured in another media form. Some of the higher profile exposure had come from a couple of earlier TV shows, one of which was aired by the History Channel back in January of 1979 that was narrated by the late Leonard Nimoy of Star Trek fame. That show was titled *In Search Of*, a series of documentaries that was shown on the Sci-Fi Channel, a branch of the History Channel that were aired from 1977 to 1982. Oak Island was featured in episode #16 that highlighted the ongoing search for the mysterious cursed treasure. The History Channel in 2002 started to re-air episodes of that series of documentaries along with others for a few more years. In 2013, when the current producer of the television series for the History Channel shows approached the Lagina brothers and their partners about doing a new series of their endeavors on Oak Island, one would think they would be elated, but at first were hesitant.

The chance to have a TV series featuring what the Lagina partners would be doing on the island would, after some reluctance, be accepted though. Since January 2014 the show titled *The Curse of Oak Island* has been aired regularly each year captivating audiences worldwide with viewer participation in the millions. The search efforts shown on the show have presented to the audiences several successes with the discovery of many important artifacts. A few of those finds have helped to narrow one of the most nagging questions as to when events occurred on the island.

In 2015 when the TV show was in its infancy, Dan Blankenship wanted the Lagina brothers to try once more to investigate his biggest challenge of exploring Borehole 10X. This renewed attempt long awaited by Dan would be one of the new series episodes. All involved agreed to first send down cameras to examine if it was safe to enter into the murky water filled shaft before thinking of risking any dives by humans. With TV viewers now seeing what was occurring on the island, they were captivated by the scenes of the equipment surrounding the area of 10X and the preparation of underwater divers readying themselves for a dive into the cylindrical metal shaft. I believe the entire event was aired in several episodes that covered four dives over a period of two days. Those first airings of *The Curse of Oak Island* were filled with intrigue and danger that would attract more viewer interest for years to come.

Since the years of the aired TV show, the Lagina team has conducted more than forty search drillings recovering what has been determined to be several samples of wood that were thought to be used in the construction of the original Money Pit. These samples of wood have been dated using dendrochronology to be around 800 years old! Since the Money Pit was discovered in 1795, it would put the possible construction date of the pit as having occurred around the beginning years of the 1400s! There is a lot of history that can be theorized during that era as to who might be the originator of the events on the island. That is only one part though of this complex mystery. One very odd discovery that has occurred since the Lagina brothers have been retrieving items from the digs, has been the retrieval of two ancient human bones! The bones were radio carbon dated to be from around the 1600s with each bone piece having a different DNA hereditary signature indicating that each had come from different geographical origins! One bone piece was determined to be Middle Eastern, while the other was more European! I have a good theory about this that later will be explained.

There have been other very interesting items discovered that can help tell of possible time periods of activities occurring on the island. A member that was added to the Lagina team after a few years was the introduction of a very successful treasure hunter named Gary Drayton, a metal detection expert. Drayton is originally from England where he began as a hobbyist searching trash pits for any thing of value, especially old bottles, finding some that dated back hundreds of years. Gary would move to Florida where he continued his searching hobby in the local swamps and surrounding beach areas. In Florida, he found some old Spanish gold and a emerald encased ring worth around $500,000! His notoriety about his finds were noticed by the Laginas who invited him to use his talent to scour Oak Island. Drayton's searches have discovered all sorts of relics so far, they have included handmade iron nails, ships' spikes, old military buttons, various aged metal digging tools, a silver Spanish coin, cutlery, oxen metal shoes, pottery, a brooch pin with garnet stones thought to be Victorian, and many other things. All of these findings by Drayton are valuable in identifying historical significances to previous Oak Island activity.

The biggest find of importance so far in my opinion by the Lagina team has been the discovery of a crosslike shape made of lead metal that was found at Smith's Cove beach in 2017 while filming Season Five. Rick Lagina and Drayton were searching the beach area during low tide when Drayton heard his metal detector make a distinct sound indicating something metallic. What they found that day was a piece of incredible history in the form of a unusually shaped small lead cross. Rick and Drayton knew immediately that they had discovered something that had never been seen before in North America! The small cross which measured about three to four inches vertically had a cross bar of about two to three inches horizontally that is identified as a Tanit. That symbol is known to represent an ancient female goddess that was worshipped in the ancient regions of the eastern Mediterranean, and more precisely in the region known as Phoenicia. After studying the cross for a moment, Rick realized that he had seen this cross before on an earlier trip to France! Rick had seen the same shape of the cross in a Knights Templar prison in Domme France where it had been carved into a wall of the prison! Domme had been built as a fortified town around the year of 1281 that sat atop a hill in the Dordogne Valley region of southern France. Local historians had noted that some Knights Templar had been imprisoned there after they had been arrested in 1307 by orders of the French King Phillip. That date of Friday the 13th 1307, is known by most as Black Friday for that reason! The cross was later analyzed by scientist who dated the lead material as to have been made between 1200 to 1400 A.D. possibly coming from lead mines once found in eastern France! A special note here that the area of southeast France during that time period of this lead cross dating is also the same region of the Knights Templar beginnings! This cross is also thought to be similar to that of another ancient idol known as Ankh, an ancient Egyptian symbol also having similar sacred meanings. In an aired video conference with late author Zena Halpern during Season Four, she describes to the Lagina team that the symbol is an ancient Tanit. Both symbols are very similar with the Tanit originating also from the Middle Eastern lands of Carthage or Phoenicia. Both of these symbols represent the worship of a female goddess. The lead cross found on Oak Island leads one to believe that a visitor to the island must have had a similar belief in this symbol. Ironically, the historical places around the Mediterranean where the Tanit was worshipped are also some of the regions of Knights Templar activity! More about this later! This is an extremely puzzling find as it does present the question of where did the cross come from and who was the possible owner, or in this case who was the loser of the cross? Anyone caring to speculate that the cross must have been lost by a Knights Templar could be correct!

An illustration by the author of the lead cross or Tanit found on Oak Island

Tanit or Ankh symbol

One more location that may have similar wall engravings of the Goddess Tanit can be found in a cave located in the town of Royston, England found just north of London. This location was probably used as a safe shelter, or more likely as a secret lodge location for the Knights that were trying to avoid capture after 1307. Accepting that not all of the Templar had been arrested in France, and some had escaped King Phillip's soldiers, a good place to hide for a time was in England. It was far enough away from France to avoid capture.

The cave was shown as part of the historical television series titled *Buried Knights Templar and The Holy Grail*, which aired on the History Channel in 2018. The documentary shows investigators Garth Baldwin and Mickey Kay tracing the possible journey of the Knights Templar from Jerusalem to Nova Scotia. Garth Baldwin is an archeologist while Mickey Kay was a former British soldier who was a military strategist. Together the duo analyzes the knights activities during their 200 year history. One segment of their journey finds them inside an old small brick and stone building in the town of Royston, England where the duo are led down a secret series of steps to a cave like location by its curator. The cave appears to have rounded walls made of stone and mortar with hundreds of carvings of various sizes and shapes upon the walls. The largest of a number of shapes appear to have some resemblance to a Tanit. When this documentary was filmed, Kay and Baldwin were probably not aware of the finding of the Tanit on Oak Island. Their documentary was not released for showing until January 31, 2018 with it being in the production stages long before that date. The Oak Island Tanit find had just occurred a few months before in the summer of 2017 meaning that it would not be aired until after the release of their documentary. The *Buried* documentary is presented in a four-part series with historical commentary presented by the world renown author and expert of the Knights Templar, Dan Jones. All is done very well with a suggested viewing of the production as a learning experience for those interested in the Knights Templar.

The proof of a Tanit found in those two locations known to have had a Templar presence surely gives credence to the thought that a Templar lost the Tanit found on Oak Island!

The television series about Oak Island has been very entertaining to say the least, always leaving the viewer with an appetite to see what the next episode will discover or attempt. The Lagina team has added more researchers and experts and have now cloned themselves as the Brotherhood of the Dig. They have continued to drill many more bore holes in search attempts to locate the original Money Pit. They have also utilized ground penetrating magnetometer technology to do surveys of almost the entire island creating a comprehensive mapping of what lies under the surface of the ground. What has been discovered is that there are, as was suspected in the area of the

Money Pit, many different voids in various directions that connect to several tunnels, shafts, or chambers. They also have discovered during the GPM surveys, that there were several anomalies in other areas of the island that indicate ground disturbances that are possibly evidence of where something may have been buried. All of the drill probes into the areas thought to be where the Money Pit might be located, have been tireless efforts to narrow down where a new dig should be done. That is what Dan Blankenship had hoped Borehole 10X would accomplish, locate the original Money Pit, or at least be able to close off the source of the flood tunnels.

While there has been a focus on the area of the Money Pit, there has also been a simultaneous effort to discover what secret the swamp holds. This piece of triangular shaped lowland that seems to be created between the two larger land masses of the island has been long suspected of once being a well used ships' harbor. The team, led by Rick Lagina, have uncovered many finds in the form of what appear to be several short pieces of wooden ship planking, hand railing, small metal fastening devices, possibly from old chests or boxes, and other items that were possibly used on very old maritime sailing vessels. When watching the History Channel TV series, the viewer often sees Rick wearing muddied jeans and casual attire with shovel in hand prying the swamp spoils that Billie Gerhardt has excavated with his machine. They both watch Gary Drayton scan the dredged sludge of the swamp with his metal detector hopefully waiting for the sound of a discovery. There has been a suspicion for many decades that the sunken remains of an actual large sailing ship as being buried in the swamp. The Lagina team has employed several expensive resources in efforts to identify what is there as another part of their quest to understand all of what the story of Oak Island could reveal. One such tool employed in their search efforts of the swamp has been the use of a sophisticated type of electronic seismic recorder. The recorder uses small explosive devices placed in a well designed grid that maps the underground with echo readings from the discharge of the explosives. What those recordings have revealed is that there is some type of an anomaly buried in the swamp shaped similar to the hull of a large ship. The object has not been uncovered nor identified as of 2021.

The swamp has also disclosed that at one time it was the possible site of an offshore ships' landing wharf, or wooden pier that had access to a stone road at the edge of the swamp! The road was found by using good old manual labor with hand shovels, plus Billy on his excavator. The roadway seems to lead from the water's edge of the swamp toward the direction of the Money Pit! The stone road, which is about eight-feet wide, seems to be very old and was created by using the beach stones found on the island. The Oak Island shoreline is littered with 10 to 20 pound size rocks, just right for a good road foundation. The road is believed to have been used for carts carrying whatever, or as a walkway from an anchored ship or skiff to the area of the inner island. The roadway has been estimated to have been done long before the time of the Money Pit find!

Fred Nolan had always believed that the swamp was hiding something of great importance related to the events having occurred on the island. During his many surveys of that area, he would often find what was described as being survey stakes that were made from wooden poles, finding as many as dozens of them. In Fred Nolan's mind, they had to be used as references to locate something, or to position it properly on the island then probably to record it on a map. This seems to be another puzzling piece to that discovery because there is no recordation of where Fred Nolan found the stakes that were later discarded. He may not have wanted to reveal what he believed they were referencing. Fred Nolan was somewhat of a secretive man when it came to disclosing what he had done on the island out of mere caution to protect what he knew. Fred was a smart man who had been in a constant feud with the other owners on the island that had led to costly court time and expensive lawyer fees. He owned the swamp area, and his business about what he did was his to protect!

The latest venture in 2021 that the Lagina team is seeking to attempt is the excavation of a suspicious bore hole that has been found to be the possible site of the original Money Pit, or a flood tunnel that could lead to the pit. This bore hole in the area of the once known Money Pit has been chosen through the process of elimination after several years of searching that has lead to dozens of search holes being drilled. All of the search results have been carefully analyzed and mapped creating a trail to this one particular site. This particular bore hole has yielded evidence that is believed to be the remains of wood used in the original money pit shaft. Electronic sampling done in the depths of this hole have also indicated the presence of a possible large quantity of silver! What will be done next is to excavate the shaft hole with a very large type of oscillating auger several feet in diameter that removes dirt after digging downward. As the hole is cleared it will allow the insertion of several huge 10 foot in diameter round steel cylinders to be placed end to end into the hole. This is a very similar effort done by Dan Blankenship in the past with 10X, but on a spectacularly larger scale. Could this possibly lead to the victory that has been sought for more than 200 years, the discovery of the actual Money Pit treasure? As I write this book in early 2022 there is no answer yet.

I would suggest to anyone reading this book if you haven't seen the TV Oak Island series by now, you should try to view as many re-runs as possible. I would imagine anyone reading this book has already seen their share of the series' shows which I hope are still being aired. How long will the Lagina team search is anyone's guess at this time.

It is time to leave this part of the book to now enter into the who, what, when, and why that will finally give the answer to those questions that lead to the truth about the Oak Island Mystery!

Dan Blankenship remained dedicated to his search efforts as well as his continuation of the feud with Fred Nolan that had lasted almost 50 years. Dan however would never get to discover what the Oak Island treasure is as he passed away in March 2019 at the age of 95.

Fred Nolan kept doing his investigations of the swamp while keeping any beliefs of what he thought the island was hiding as knowledge he would carry to the grave. Fred Nolan died in June 2016 at the age of 89.

Both men had spent more years searching the island than any one else had ever attempted, leaving a legacy for their heirs to solve, the mystery of Oak Island.

Who Did It?

So the discovery of something on Oak Island has led people to believe that there is a treasure of great value buried in a deep pit that has dared many to attempt its recovery for over 200 years. How did that idea formulate of a treasure being buried on this particular island? Maybe the idea or legend grew out of the mere imagination of what some believed could have happened on the island. Call it suspicious minds, call it fantasy, call it story telling lies, or is there some truth about it all that we just can't verify! Is there a treasure buried there? Who says so? Can they prove it? What is it? No on knows! There is no real proof of anyone ever burying a treasure there! Think about that for a moment or two!

Oak Island is a true mystery because the story has no answers. At least not yet!

In trying to understand what the Oak Island mystery is all about, one must examine not only the history of the events since its beginning in 1795, as these first few chapters of the book have done, but to pursue a thorough investigation that seeks to answer the who, what, when, and why.

Without answering those particular questions, there will never be any relief that leads to the truth about the mystery! Only those answers can help in knowing the truth if something of great discovery is truly hidden. For centuries the most common belief is that the mystery of Oak Island was the result of the actions by a few named pirates, or a band of countrymen from France or England, or perhaps even Scotland. Maybe even those of a military order on a mission to secret the crown jewels of a kingdom. Who knows what others it could be! The suspects have been numerous to say the least. There is much to sift through, so let's begin!

The place to start logically is who it would not be. I cannot imagine that a ship full of so-called pirates as being the prime suspects! Why would they even consider the idea of constructing such an elaborate scheme of the Money Pit to secretly hide treasure. That concept would probably be laughed at by almost anyone, especially a pirate. Not to say all pirates may have been ruthless and heartless individuals with evil self interests and having any abilities, but this was not the pirate way of doing things. (More on pirates to come.)

So who else? There were many explorations to the New World by the French and English and other European adventurers prior to the notable years of the Golden Age of Piracy. The North American lands and their plentiful resources were ripe for the taking. A new age of discovery and exploration had begun in the 1500s which would last until the 1700s. Europe was expanding, globalization was beginning, and any new territory that could offer valuable new resources for the European continents' growing population would be a prize worth the risk of the adventure. Fortunes could be made with the discovery of any new resource that was of value, like gold, silver, iron, even a new food crop such as the abundance of fish, or the possible trade in furs with the indigenous Mi'kmaq people. In the pursuit by each European country to expand their empire, the North American continent was awaiting those willing to venture forward into the unknown perils of the vast territories of the unexplored far north, south and west. The influx of explorers searching the coastlines and inlands of North America were in constant conflict with each other. They all were trying to be first in claiming ownership of the riches found in the new continent with the bountiful lands across the sea.

Christopher Columbus, the most noted explorer, had already discovered part of the New World in 1492, although he was not the first to set foot upon the lands of the far north! That honor as history proclaims, goes to Leif Erikson (circa 970-1020), a Viking who actually established a settlement presumably on Anticosti Island in the St. Lawrence Bay area of Canada. Leif Erikson however did not claim any territory for his homeland of Greenland where he would return and remain until his death.

Even before Leif Erikson had sailed to the shores of North America, there is an account of a 6th-century monk named St. Brendan who would sail from the shores of Ireland toward North America on one of his ventures in search of his Eden before returning home. He was known as The Voyager. Neither of these individuals leave any evidence of ever venturing onto Oak Island and in all probability did not even know about its existence. Even if they did know, what reason would they have had to come to that part of the unknown world to construct anything on an island, especially one that was a deep pit, or a collection of huge boulders that resembled a huge Latin style cross like the one Fred Nolan discovered?

It wasn't long after Christopher Columbus had sailed the route from Europe to the New World, actually repeating the trip four times, that other Europeans would begin seeking the new continents resources. As early as around the 1520s, fishermen from the more northern areas of Europe had begun sailing across the North Atlantic to the rich fishing grounds of Nova Scotia. They pursued such plentiful fish as cod, salmon, striped bass, tuna, trout, and others. The harvested fish would be salted down for the trip back home which usually took the sailing ships about 6 weeks with good weather to complete. In those days, salting was the only means of preserving foods as refrigeration had not yet been available until the late 1800s. Salting would keep foods edible for as long as a year. Even though a seasonal fishing camp or two could be found in and around Nova Scotia, the large number of ships fishing the area were not, in my opinion, suspects as to what happened on Oak Island!

What would be more likely are those who would come later, like the French and English Explorers along with the possibility of a few of the Portuguese and Italians that were representing their countries as being the most suspect of doing anything eventful on Oak Island.

There are some names many will recognize that could be suspect especially after the discovery by Christopher Columbus of the eastern coast of North America in 1492. One very recognizable Italian name known to those around New York is Giovanni da Verrazzano (1485-1528), who sailed into New York harbor in 1524, thus the Verrazzano Bridge named in his honor. Another very famous Italian name was that of Amerigo Vespucci who first discovered Rio de Janeiro in 1497. His name would be used to identify the northern and southern continents, America. Other explorers like the Portuguese or Spanish such as Vasco da Gama (1468-1524) who sailed to India, or Ferdinand Magellan (1480-1521 who found the route around the southern end of South America to the Pacific Ocean via what is now known as the "Strait of Magellan."

In 1497, the Italian explorer named Giovanni Cabot, also known as John Cabot by most, sailed from Bristol, England with a crew of 18 heading northwest toward the new continent. Financed by the Italian banking establishments, but sailing for the King of England, he landed in Nova Scotia somewhere around Cape Brenton in the northeastern part of the peninsula claiming that area for the King of England before returning home. Cabot made a second voyage in 1498 with five ships and 200 men. On his second trip, he was accompanied by a group of Catholic friars which are believed to have stayed and established the first Christian mission under the authority of the Roman Catholic Church north of Nova Scotia in Newfoundland. This was the first formal introduction of Christianity in North America.

Spanish explorers such as Cortez and others like him became mostly interested in Central and southern South America, which required a different sailing route from Europe than those going to North America. The Spaniards used this route farther to the south to avoid the more northern fleets of the French and English who would be hostile toward them probably sinking the Spanish vessels upon sighting!

There were several other explorers during the years of the 1500s to the 1600s, but they were seeking other lands of the north during this time period as to warrant any suspicion regarding Oak Island history. Jacques Cartier (1491-1557) of France mapped the Gulf of Saint Lawrence, discovered Prince Edward Island, and explored the west coast of Newfoundland. Samuel de Champlain (1603) sailed the eastern coast of North America from Mexico to Newfoundland and established Quebec City in Canada known as Nouvelle France (New France).

It wasn't until around 1604 when the far northwestern part of Nova Scotia was the site of the first attempt at a settlement by 79 French colonists led by the explorer Pierre Dugua. They would make an encampment on the island of Saint Croix. The expedition to establish a colony on Saint Croix was short-lived due to their inexperience in dealing with the brutal winter weather of the far north which was even further complicated with them suffering an outbreak of illness known as scurvy. Scurvy is the result of a diet poor in nourishment, usually resulting from the lack of vitamin C caused by insufficient ingestion of fruits and vegetables in the human diet. About half of the colonists that landed during that first encampment either died from the scurvy, or exposure to the brutal weather. Pierre Dugua along with Samuel de Champlain his cartographer, and Jean de Biencourt moved the remaining colonists the following year to their new location father to the south and named it Port Royal. Port Royal is located on the western side of Nova Scotia to the south of the Bay of Fundy and about midway of the island. Sieur de Mons who led the first landing in 1604, returned to France in 1605.

Port Royal, so named for being a sheltered harbor offering safety for ships, had now become the first established European colony north of St. Augustine, Florida which was established in 1565 by the Spanish. Nova Scotia was beginning to transform from a seasonal fishing destination for Europeans into becoming a new frontier for the French which they referred to as Acadia, meaning "New France." Even though the French had claimed territory on Nova Scotia, the settlements were very small with a limited number of colonists.

After only two years, Port Royal would be abandoned in 1607 as it was not worth the investment by the French government then led by King Henry IV. The most valuable resource the Acadians were sending back to France to help justify the existence of the colony were the harvested pelts of the local fur bearing animals. Evidently it wasn't sufficient enough to sustain Port Royal!

An early attempt by the British in 1585 to colonize the Mid-Atlantic coast of North America had landed an expedition in mid-June on Roanoke Island, North Carolina. Queen Elizabeth I of England had a need to expand her realm due to the population growth of the country and needed more resources that were vital to fulfill the needs of her people. The Queen also wanted to claim some of the New Continent for England before the Spanish acquired them, or before the French could expand their occupation that could hamper any English exploration.

Queen Elizabeth would sanction Sir Walter Raleigh to explore and colonize the lands of the new continent to the west of Europe. Raleigh was to oversee the establishment of a colony and a base of operations for English privateers. These colonies would be within sailing distance of the Spanish treasure ships that were laden with silver treasure taken from the South American territories. Raleigh, had authorized Sir Richard Grenville to lead the mission and Ralph Lane to be governor of the new land. Lane, with about 108 colonists came ashore to begin a colony on Roanoke in the summer of 1585. After the landing on Roanoke Island an encampment had been established. Sir Richard Grenville left a small group of colonists at the settlement along with Lane as their Governor, then returned to England. Grenville's return to England was to prepare for another return trip to the island accompanied by more colonists and to re-supply the colony. John White was appointed by Sir Raleigh to lead a second wave of colonists to Roanoke Island in 1587. Upon White's arrival, he found the settlement vacant, all but abandoned except for a few remaining living quarters in disrepair. The total disappearance of more than 100 colonists within a year is still an unsolved mysterious event. Sir Walter Raleigh will be mentioned again as he is involved in a very mysterious situation about Oak Island.

It is recorded that Sir Francis Drake, another of Queen Elizabeth I representatives, had sailed to Roanoke to check on Lane's settlement around mid-June 1586 that resulted with his rescuing some colonists before his return voyage to England. Sir Drake had just finished his successful raids on the treasure laden Spanish fleets near Florida in the waters of the Caribbean. Spain and England had been at war on the high seas for decades which is known as the Anglo-Spanish War until its end in 1604. Sir Drake's ships which had numbered as many as twenty-one on this encounter returned to England on July 22, 1586.

Jamestown, Virginia also was to become an English settlement in 1607 with the occupation by 104 men of various ages. That first year saw the completion of a fort to protect the settlers from the local

Indians and any Spanish who may invade. Jamestown however did not become a success at first. Between 1609 and early 1610, almost all of the settlers had died from starvation. They had been confined to their fort in fear of the local Powhatan natives and were unable to forage for food. The resupply ships from England had failed to keep their schedule due to being stranded in Bermuda because of ship misfortunes and repairs. It would not be until May 1610 before the resupply group from Bermuda, now sailing in new boats constructed in Bermuda, would finally arrive to save the remaining settlers.

Port Royal, the French settlement in Nova Scotia, unfortunately was burnt to the ground in 1613 by a group of British coming from the lower parts of the mainland continent of America. The French would later regain control of the area rebuilding the fort once again. It was often during this era that a warring foe traveling the coast would invade any new settlement held by another nation in an attempt to dissuade their establishment and to claim the territory they occupied. Sometimes the settlers escaped inland to avoid capture or death, or would even pay a ransom to the invaders in exchange of not being harmed. It was not unusual to hear of a settlement being attacked by an opposing country that led to it being completely destroyed and the residents never to be found afterwards. It wasn't always pirates or privateers who were the enemy of those seeking to form a colony or settlement, it could be other foreign colonists or even the local native tribes that often were the threat.

In 1629, the Scottish arrived in Nova Scotia to establish a colony claiming ownership of all the lands that were claimed to be Port Royal forcing the French to go farther north. Several years of war between these two colonists' groups ended with a peace agreement that eventually returned Nova Scotia to French control, but that would only be temporary.

The French would not abandon their efforts to claim territory in what they believed would become their New France, and late in the year of 1632 would establish a new settlement called LaHave on the eastern shores of Nova Scotia. This new settlement was located nearby, just south of Mahone Bay, approximately 40 km by sea from Oak Island!

There were so many battles fought over attempts at ownership of the new territory, that during a period of about 80 years, there were as many as 10 significant conflicts between the French, English, Scottish, Spanish, and Dutch. There were too many other smaller skirmishes to list here, some even with the indigenous Mi'kmaq Indians who had occupied the Acadian area for hundreds of years prior to any foreign intrusion. There are numerous books and articles filled with those details about the many skirmishes and wars during that part of history – enough to keep a reader engaged for years.

1713 found the French military constructing a fortress at what would be called Louisbourg on the northeastern shores of Nova Scotia. The area was the former landing site in 1497 by the Italian explorer John Cabot who had sailed for the English crown. It was known as the area of Brenton. This area was used by many European mariners, especially those friendly to the English. In 1745 the British laid siege to the fort burning it to the ground. This area would change hands back and forth a few more times throughout history.

During those decades of back and forth conflicts, and settlement failures, there could have been an occasion when ships were seen around Mahone Bay where Oak Island is located, especially after the Protestant establishment of Halifax in 1749 which was just 60 miles to the north with the largest naturally sheltered harbor able to protect ships in the north east region. Halifax would soon become an active British military stronghold, and later was to become the main population center of eastern Nova Scotia.

Nova Scotia and the surrounding land areas were consistently being invaded by one country or another, all competing for any valuable natural resources they could find with each claiming the territory unto themselves.

Chester, which is south of Halifax and just a few miles north of Oak Island, wasn't settled until 1761 by the means of a land grant from England. The French Acadians had been forced out of the area once more, and immigrants from the lower continental English colonies were allocated the free land to establish an English presence.

Most competing land grab attempts around Nova Scotia had led to the various conflicts and wars between nations came to an end in 1763 with the signing of the Treaty with Paris. That treaty gave the British, control of almost all of the French Acadia lands. Those lands reached from the north of Nova Scotia, southward all the way to Kennebec River in southern Maine. French exploration and territorial claims eventually shifted to the region around the Great Lakes and to the northern territories of Canada. The French language still remains very evident in parts of Canada today.

All of these wars and conflicts were attempts toward the early colonization of Nova Scotia and the other nearby northern territories, and resulted in much daily turmoil and hardship for those trying to establish a new land upon which to survive. The constant threat of danger from enemies, plus the hardships presented by the new wilderness itself, left no time for the very few colonists present to be involved in anything other than acquiring their daily necessities for survival.

I find it unlikely that the adventurous colonists who existed upon the shores from Virginia to the northern waters of Nova Scotia during this time period, would have had any involvement in doing anything so complex that we now know as the mysterious Oak Island!

What hopefully has been presented so far is to give an indication there was a very limited permanent presence of European inhabitants in the area of Nova Scotia, especially Oak Island! If there had been any witnesses to the events on Oak Island, it is nowhere written! Why?

Caveat! I do find it plausible that there is the possibility of one or more of the warring nations' ships, or their privateers, could have buried coinage and other valuables on Oak Island. It is possible they wanted to avoid seizure of their treasury by any opposing forces should they succumb to defeat or capture! They could have stashed their treasury for safe keeping either during or before conflicts or exploration, which would be a very logical strategy, also very smart. A crew of privateers aboard a war ship could easily have had enough capable manpower to hide their valuables if such was thought to be necessary. Could there really be some treasure buried on Oak Island? Why not! It has been told that one of the original Oak Island trio paid for some goods once using foreign gold coins! There is another story of three treasure chests presumably buried on the island that have never been accounted for to this day! There are the facts about a free black farmer being granted land on the island who would eventually buy more before his death. Some have said he made those purchases using gold!

But we still have not discovered the answer to our first question, who? It is time to look at some more suspects. So, who is next?

The Suspects

It has been a little more than 200 years since the search for a possible treasure on Oak Island had begun, and the one big question that still remains, who was the entity that put it there, and why?

There has been much speculation as to who did what on the island. Over the past centuries a list of suspects has been developed ranging from pirates to privateers, possibly some colonists, maybe even a governing body of the new territory. There has even been suspicion about the legendary Knights Templar who may have traveled from as far away as the Holy Lands of the Middle East. All have been mentioned in lore, either from the storytelling by word of mouth, or in print found in the many books and articles that have been written over the decades since the legend of Oak Island had begun. It has not been an easy task to try to identify a time frame when such an occurrence actually happened, as the time in history for each suspect is different, spanning from decades apart, to many centuries. We do however know that what initially occurred on Oak Island, happened prior to 1795, the year when the suspicious Money Pit depression was discovered by the trio of Smith, Vaughn, and McGinnis. With each of the possible suspects being active in a different time period in history, there is a need to carefully analyze if there is any proof that could link any one of them as being the prime suspect. They not only must have the potential for any reasonable involvement, but the opportunity and a meaningful purpose. For example, during the Golden Age of Piracy, which was described as the most active period of piracy from 1650s to 1720s, numerous pirates were known to roam not only the Caribbean, but also the North American coastline. There have been some stories about the possibility of a few of those pirates active during that era who could be considered as suspects involved in the events that created Oak Island.

In an earlier chapter, I described a few of the first explorers and colonists that had attempted settlements of the new lands of North America and Nova Scotia, mostly in the time frame prior to the Golden Age of Piracy. But, there were a small number of others known as privateers who were also active during that same time. Remembering that a privateer was like a pirate, who operated under the license of a sponsoring country to seize the ships and cargos of non-allies. These privateers were like a naval military and did not only attack other nations' ships to control the seas and lands, but also to confiscate any of the valuables to enrich their own sponsoring country and themselves.

First, I want to acknowledge who the Lagina brothers and their partners have suggested from their own findings as some are good suspects. None of their research so far though has been able to definitively prove the involvement of any particular person or group. It is not that they haven't been trying, quite the opposite. They have had numerous researchers and historians aiding them with every possible aspect of their quest. Every little piece of information relating to the island and its history has been extensively researched, scrutinized, evaluated, and documented. Every item that has been discovered resulting from their numerous diggings and drillings has been thoroughly analyzed and many scientifically assessed. They have traveled to Europe, visiting some of the historical sites referenced as possible leads as to those who may be involved in the creation of the lore surrounding Oak Island. Many hours have been devoted in talking to theorists with different views about the possible connections that some suspects may have. It is also without question that this pursuit by the brothers, Marty and Rick Lagina, partner Craig Tester, plus the television producers of the History Channel, has been exhaustive, time-consuming, frustrating, and expensive. Yet, there has been no conclusive answer as to who, why, when, or what!

William Kidd

The first of several suspects to talk about is the infamous Captain William Kidd, who lived from 1645 to 1701. Kidd seems to have garnered a good amount of attention as being a prime suspect for depositing hidden treasure upon Oak Island. I believe this is because he was more well known from once living in the New World area of New York. Rumors about the deeds of well known people can sometimes become distorted into popular pirate legend.

Kidd was born in Dundee, Scotland and early in life his family had to rely on charity after the death of his father. Dundee was located on the northeastern coastline of what is now the United Kingdom. During the 17th century, Dundee was mostly known for its export trade of dyed wool and hides, and for the importing of wine from France and Spain. The area population during this time was approximately 11,000 with much of the citizenry being in the wool business, or as merchant sailors and fishermen. Dundee was under constant threat by the English as the two countries continually warred, mostly about their religious differences and territory. In the years around the 1790s, the port began to lose its importance due to the constant armed intrusions by the English which interfered with much of its trade activities around the North Sea area.

Kidd was around the age of 25 when he left Scotland. Earlier in his life, he had served in the Navy during the Dutch wars of 1673 as an apprentice seaman. After serving in the Navy, he became a merchant seaman with his own ship, but eventually would end up aboard a pirate ship under the command of a Captain Jean Fantin. In the summer of 1689 while sailing aboard the pirate ship named *Le Trompeuse* with a mixed crew of French and English seamen in the Caribbean, Kidd took part in a mutiny aboard the ship. The captain was ousted and the crew set sail for the British colony island of Nevis located south of the Dominican Republic. Kidd, who had plenty of naval experience, was named captain of the ship which he promptly re-named the *Blessed William*. Since the island was under the control of an Englishman, and England and France were at war, the island's governor, Christopher Codrington, recruited Kidd and his crew to protect the island from any French hostilities. This is the beginning for Kidd to be recognized more as a privateer than as a pirate by his deeds at sea. Nevis was a safe temporary harbor for Kidd, and although the governor could not pay him for protecting the island, it was a base of operations from which Kidd used to launch his own forays of the other nearby islands. One of the islands Kidd did plunder was one which was being occupied by the French. The Kidd crew would destroy its only community, and enjoyed a booty of an estimated 2,000 pounds sterling.

Unfortunately for Kidd, after the pillaging of the French-held island Marie-Galante, some of his crew were not pleased with the tactic of inland combat, and one day while he was ashore on the island of Antigua, some of the crew mutinied. They then sailed away with the *Blessed William*. Kidd, still in the favor of Governor Codrington, was given another ship that had been previously captured from the French, and was appropriately given the name, the *Antigua*.

New York was an attractive destination during this time period, established as an English colony in 1665, it had been previously known as New Amsterdam under an earlier occupation by the Dutch. The city had a great harbor and was centrally located on the east coast of North America. The area being almost totally controlled by the English offered refuge from the war ships of the French armadas sailing in the waters to the south. Kidd decided on safety and sailed the Antigua northward into New York harbor around late 1690 or early 1691, just missing the departure of his former ship, the *Blessed William*. New York harbor had become a haven for pirates and privateers of the English fleets as there were many merchants located in this trading hub willing to barter anything profitable, even if it was ill-gotten plunder.

This was a good fit for Captain Kidd who did not really have a reputation yet as being a pirate, but more that of a privateer. It didn't take long for Kidd to acquaint himself with the leadership in the New York area, coming into favor with a local governor or two, and marrying a wealthy prominent young widow. Now married and socially well known as a very capable seaman, who had been a formidable privateer once helping the governor of Nevis, he was offered a proposal to privateer in the name of the crown on behalf of the English governor Bellomont. Bellomont controlled the three states of New York, Massachusetts, and New Hampshire. The war with France had offered plenty of opportunity for Kidd to continue to be a privateer and to gain more wealth, but being married to the very wealthy widow Sarah Oort, he really did not need to continue that pursuit. But in 1691 he would accept the offer to privateer in the name of the crown, capturing a French merchant ship off the coast of New York which earned him a reward from governor Bellomont. What would later become a misstep however for Kidd was that during another incident of so-called privateering, Kidd boarded a favored English ship and seized supplies for his own ship. That act, deemed one of piracy, would cost him a political reprimand from the governor that would cause the loss of his privateering credentials, a black mark that would eventually come to haunt him in the eyes of Governor Bellomont.

In 1695, Kidd however would receive another command to captain the sailing ship *Arigo* that was assigned to deliver cargo to England. While in England he was approached by a friend from New York who had come to London to testify before the King's court about some political mischief back in the colonies. During this trip, it was learned that the King was unhappy about the amount of piracy the country was suffering, especially in the area of the East Indian Ocean. England had been trading in spices, silk, tea, and slavery beginning around 1600 when it established the East Indies Trading Company located in Mughal, India. Robert Livingston, who was the friend testifying from New York, received information about an anti-piracy venture to thwart the problem in the Indian Ocean and suggested Kidd to become the task commander. King William III of England agreed, and would provide a letter of marque, which was a royal license to privateer on behalf of England. The Crown would receive 10% of any bounty taken, leaving Kidd and his crew to split the remaining fortunes. Still sailing the *Antiqua*, the ship would not be adequate enough to fight some of the French warships encountered, so the *Antiqua* was sold to help purchase a new 124 foot long, 34 cannon, 150 man crew ship named the *Adventure Galley*.

This venture would eventually lead to the demise of Kidd. On April 23, 1696, Kidd set sail from England with a partial crew of 70 heading back to New York for more crewmen. But before Kidd could clear the Thames River, a English naval military party boarded his ship drafting some of his crew, then exchanged them with others of character not so desirable. While heading back across the Atlantic to New York, the *Adventure Galley* scored its first prize, seizing a French fishing ship bound for Newfoundland and sold it in New York. This would be one of Kidd's last legal bounty trophies as a privateer. Upon arrival in New York, Kidd could only find a limited number of able-bodied seamen who were willing to sail to India. The replacement crew was less than desirable in character, but were necessarily mustered anyhow as sail was set for the Indian Ocean.

Kidd would reach the Comoro Islands located off the far southern eastern shores of Africa in February 1697. The search to find pirate ships would began without any early success. The Comoro Islands lay between the east coast of Africa, but to the west of the island country of Madagascar, the largest island country located in the Indian Ocean. This was like a funnel for sailing ships with the Comoro Islands situated midway between the two land masses, it was an excellent location for a base of operations. Unfortunatley for the ship's crew, a cholera outbreak would cause the loss of about 50 men. Now, shorthanded again, Kidd was forced to capture a number of the Comoro Islands locals for the ships duties. The entire crew became an odd mix of skill levels, ethnicities, ages, and languages. The crew was becoming displeased with their uncertain situation as they had hoped to prosper from much promised plunder, but instead were left with a lot of time to complain. Mutiny would be rumored as a possibility.

Summer of 1697 was to give the crew some hope when seeing another set of sails approaching, but it would only be one more of the friendly East India Company's ships, and, as before, Kidd could not attack. He also had held back on attacking other ships as well, either being Dutch merchants, or other colonial privateers, as these ships were not the ones he had license to attack. Disappointment would began to lead to anger with the captain; eventually he did attack two Portuguese ships, whether or not they were the targeted pirates, or just trade merchant ships is not known. The hostile emotions festering among the crew erupted upon their captives of the two ships, whipping them to disclose where valuables were possibly hidden aboard. Their actions would only be a temporary release for the frustrated crew. An occasional need to re-supply the ship, would require the crew to come ashore on the local islands for water and food. During one such landing, while on the island of Laccadive, the misfit crew, on the edge of mutiny, committed a rage of attacks upon the locals. The crew abused the natives, raped their women, burned their small boats, and some of the crew even deserted. Kidd seemingly, had lost control of this rebellious bunch of misfits who were now acting more like some infamous pirates.

One more ship would be spared approach, but not without argument for inaction from one of the ship's crew named William Moore, the artillery gunner. Kidd succumbed to his own frustrations of anger over what became a vile confrontation with Moore swinging a wash bucket at him crushing Moore's head, killing him. A lull in the crews' combative attitudes assuredly waned, fearing their heads could be next.

Continuing to sail the coast in search of actual pirate ships, which was his assigned mission, Kidd sailed toward the Dutch vessel *Reparable*. Sailing under a French charter, this would be legal to privateer and Kidd impounded the ship taking its cargo of cotton, sugar, and linen goods. Not exactly the precious bounty hoped to plunder, so the ship was commandeered as the big prize renaming it the *November*.

It was now the end of January 1698, 100 years before Oak Island would be known, and the crew was somewhat less hostile, but still rumoring mutiny. They were now sailing to the east approaching the waters off the west coast of India. The *Adventure* was closing in toward the *Quedah Merchant*, an Indian flagged ship that was sailing northward toward Surat, India. This ship would become another legal seizure as the papers of the ship permitted it to sail under the authority of the French. The seized ship's cargo netted a good value in sugar, gold, silver, jewels, and silks, with an estimated worth around 70,000 pounds British. A better take than expected!.

However, there was a problem, the ship was being commanded by an English captain and technically to commandeer it would be illegal. The crew though would not let Kidd release the ship, and against the probable plea from Kidd not to do so, he took possession of the *Quedah* as a prize.

Now there was a trio of ships sailing under the command of Kidd, his original ship, the *Adventure*, the *November*, and now the *Quedah*, less its original crew. The trio sailed toward the coast of India and in April 1698 arrived at St. Mary's Islands, a group of four islands on the west coastline of Malpe, India. These islands are covered with coconut trees and are sometimes referred to as, the Coconut Islands.

Much to his surprise, Kidd found another ship anchored nearby named the *Mocha* that would even became a bigger surprise as he learned that his old nemesis, Robert Culliford, was the captain. This would be an opportune time for Kidd's revenge; it would be payback for the mutiny orchestrated by Culliford who caused the theft of Kidd's old ship, the *Blessed William*, as well as for abandoning Kidd in the Bahamas.

Kidd ordered an attack, and as luck would have it, the sortie did not go well for Kidd.

Close to one hundred of Kidd's men joined Cullifords side, burnt the ship *November*, and sailed away leaving Kidd with a skeleton crew to man the leaking *Adventure*. Only the commandeered *Quedah* would be seaworthy enough for the voyage back to England, but Kidd would have to wait for the trade winds in November to blow to the northeast in the Atlantic.

Word had traveled to the British government and the East Indian Company that Kidd's seafaring ventures representing the English crown were less that of a privateer, and now more like a pirate. His biggest mistakes were those done by his crew, plus he had killed an Englishman that was a part of his own crew. He had also plundered some ships that had proper papers that were flying the flags of friendly countries. Kidd had been labeled as just being another pirate sailing in the Indian Ocean. The tales of brutality and unsavory actions of the crew along with Kidd's own misdeeds had put him on the opposite side of the crown's wishes to end piracy in the Indian Ocean, or the world! He now had become a wanted criminal for piracy with orders given to bring him to justice. Whether all was true or not, the stories surrounding the deeds of other pirates had given the image that all were of the same lot only to be seen as the evil low life upon the seas.

Somehow, Kidd learned of this potential fate, and in an effort to get home without capture, he abandoned the ship *Quedah* near Puerto Rico which was said to have sunk after being set afire, replacing it with a smaller sloop. Early 1699, during the month of June, Kidd sailed into New York harbor sending word to his lawyer for help. Sometime while in the area Kidd was reported to have hidden a very valuable part of his remaining bounty around Long Island Sound, and possibly elsewhere.

Governor Bellomont of Boston, whom Kidd thought was an ally, would now become an adversary having Kidd arrested and put in chains to stand trial in London. Members of the crown had to save their political importance from the scrutiny as being seen in sanctioning actions considered piracy by other countries. Kidd's actions at sea, although not as cutthroat as others, were still considered those of a pirate. His pleas before the court in England fell on devious ears as there were political contentions between those in power.

Kidd may have succeeded in pleading his case that all of the contraband he plundered was legally taken, but what was inexcusable was the killing of William Moore, an Englishman. Kidd was found guilty of piracy and was sentenced to hang.

At the banks of the Thames River, at a place called Tilbury Point, London, in May 1701, Kidd was tarred from head to toe, hung by the neck until dead, placed in a gibbet, a form of human cage, and left hanging for others to see until his body eventually rotted away. Kidd's infamous body was now a stark reminder to others sailing the Thames that they better refrain from the thought of becoming a pirate.

Since over the centuries, it has been rumored that Kidd had possibly hid some of his fortune on Oak Island. Two of my big questions to answer is when and how? The number one possibility was when Kidd first came to the New York area after his time in the Caribbean. I do not believe that the engineering ability required of him or any of his crew to complete the very involved tunneling of the island, was done by them. He probably still had some of his inadequate crew of misfits that were not those with needed skills to do any engineering work as was evidence on Oak Island. My only thought of this being even remotely feasible was if he had recruited local talent after arriving in New York. It may be a possibility, but a dangerous venture to keep secret. The number two possibility is that he may have covertly sailed to the island after returning from India with a discreet skeleton crew, then actually burying a chest or two in a shallow pit as was the usual method of pirates. The finding of some coconut fiber located in a recent 2020 dig on Oak Island could be from the Coconut Islands off India, where Kidd did spend some time.

Is it all plausible? Of course, but we just can't be certain that it did or didn't happen! If anyone was going to hide something of great value, you can be certain secrecy was paramount, but since there is no verifiable real evidence of any Kidd connection to Oak Island, we can only speculate about what we don't know.

Francis Drake

Here is another potential suspect that the Laginas and one of their researchers, an historian named Paul Speed, and others have suggested may be involved. Sir Francis Drake, a pirate who became a privateer for England during the late 16th century is a good suspicious candidate in the mystery of the Oak Island legend. There is a need to investigate his history as he was a very active pirate/privateer during the 16th century under the reign of Queen Elizabeth I. This is one of the time periods from which some of the artifacts found on Oak Island have been dated, and which historical research has led the searchers to believe is relative.

Francis Drake (circa 1540-1596) first became known as an English privateer after he was commissioned by Queen Elizabeth I of England to attack and seize cargo of any Spanish ships he would encounter.

However, Drake, the eldest of 12 sons, had first begun his noted seaman's life as a pirate in 1563 at the age of 23. That is when he began to accompany his cousin, John Hawkins, who was at the time a slave trader. Hawkins had been in the illegal slave smuggling trade for years, and is thought to be one of the first to do so starting around 1560. His objective was to deliver African captives to the Spanish colonists of South America who paid him handsomely for his delivery service. Slave trading was illegal in England and the activity had to be done clandestinely. The Spanish plantation owners of the New World however needed labor, and slavery flourished making those who dared to become human trafficker merchants very wealthy. There was a rumored suspicion that even Queen Elizabeth I who came to the throne in 1558 was unaware that some of her ships flying the flag of England had been involved in Hawkins' nefarious venture. Even though slavery was illegal, there was always some form of hush-hush payment or bribe that made its way into the hands of officials and others to be quiet about the activity. Hawkins was also known on occasion to use the cannons from his ship to fire at other ships as a form of intimidation toward anyone of legal authority attempting to interfere with any of his activities. He was a pirate and a scoundrel in the worst way. Hawkins was a bad influence on his younger cousin who had become a venturous seaman with hopes to eventually command his own ship as captain.

As fate would be on Drake's side, he and his uncle would temporarily part ways in 1568 after encountering a few Spanish warships in the Caribbean. The encounter was not planned but a storm had forced them off course en route bringing slaves to the new continents of the Americas. Drake had acquired his own ship during this time that he had named the *Swan*, and as he and his uncle became engaged in overwhelming enemy cannon fire, Drake seized an opportunity to escape and sailed back to England. His uncle thought to be killed in the battle, returned a few weeks later to England to assumingly engage in a very contentious reunion between the two..

Drake was now indoctrinated into the ways of piracy and in 1572 at the age of 32, took out on his own heading from England to the northeastern coastal area of South America known as the Spanish Main. There he would plunder the treasure laden ships of the Spanish conquistadors in the area of Panama. Drake's ship, the *Swan*, was partnered this time with another ship named the *Pasha*, and it wasn't long before the two under his command wreaked havoc in the waters of the Spanish Main. The two ships were manned with a total crew of 73 and in July of 1572 captured the town of Nombre de Dios, a Spanish Catholic colony founded in 1510 on the Caribbean Sea side of Panama. Nombre de Dios was the main port used by the Spanish to export gold and other treasures back to Spain via Havana, then across the Atlantic. The English, mostly Protestant, were in constant conflict with the Roman Catholics of Spain during this era in history. In the following spring of 1573 Drake would march inland 18 miles from the port to seize a mule train traveling north from Peru. The Spanish had used this route on a regular basis carrying Spanish silver and gold; this mule train was estimated to be carrying as much as 20 tons. Drake did capture the mule train, but there was too much loot to carry back to their ships leaving much to be buried in the surrounding area of the mountainous jungle. Drake supposedly did keep a large fortune in gold. Upon his return to

England from Panama in August 1573, Drake's feat in Panama had to be concealed as Spain and England had just signed a temporary truce while Drake was at sea. The English Royalty did not want to jeopardize this new relationship they had just entered with Spain by admitting it was Drake who had seized such a valuable cargo.

Drake had become well known in England now as a young successful sea captain who had seen naval conflict. He would soon be recruited by the Earl of Essex to defend its shores from attacks at sea by Ireland. Drake was awarded three warships to command known as frigates, with which he did use successfully to protect Essex. The Earl was so pleased with Drake, that he would be introduced to the Queen as being a very adventurous and successful young asset toward furthering the goals of the crown. England wanted to conquer new territories for the expansion of its influence outside of the current realm it held. A young handsome sailing captain who just happened to catch the elder Queen's eye, favored Drake's fortune. Sometime after their acquaintance, the Queen gave Drake a small fleet of five ships that were armed with cannon to set sail for the Spanish held coastline colonies of South America and were now no longer under the protection of the earlier treaty. There, Drake was to seize what treasure he could while destroying any Spanish ships he encountered. Drake could now be referred to as a privateer, one who acts on behalf of a governing country or throne of royalty that shares in any bounties seized.

The voyage of Drake to South America began in December of 1577 leading him southward from Plymouth, England to the northern coast of Africa before crossing the Atlantic toward the eastern coastline of South America. The voyage had begun with five well armed ships and a total crew of 164 men. While sailing south off the northern coast of Africa, he captured one more ship named the *Mary*, a merchant ship of Portuguese origin. *Win some, lose some* as the saying goes, because as the fleet crossed the Atlantic, two of the ships, one titled the *Christopher*, and another titled the *Swan*, had to be scuttled due to the loss of several crewmen. It is not noted why they were lost, but sickness such as scurvy and dysentery could be a cause.

Drake's south westerly ocean course had led him to the mouth of a bay named Rio de la Plata that leads inward from the Atlantic Ocean to the port city of Buenos Aires, Argentina. Buenos Aires is located on the eastern end of South America. Sebastian Cabot, the Italian explorer, had sailed to the area in 1520 and had given the bay its name which means "River of Silver."

In the process of maneuvering the coastline plus doing some occasional onshore visits, Drake would plunder what he could of a native village, or any merchant ship that could be seized. One of his original ships, the *Mary*, that he captured off the coast of Africa, was found to have some rotting timber problems and for the safety of the crew were transferred to other ships. The *Mary* would sink into the sea. A fourth ship, named the *Elizabeth*, was sent back to England, and another, the *Marigold* was destroyed in a bad storm. Drake had lost five ships but continued to sail on alone, heading farther southward into the vast area known as Patagonia aboard his flagship once known as the *Pelican*, now renamed the *Golden Hind*. The *Golden Hind*, an impressive-looking ship of over 100 feet in length with a beam of 20 feet was a three mast baroque style ship with a large cargo capacity. The *Hind* carried 22 cannons with a fully manned crew of 80 seamen. Drake had now begun to sail around the most southern end of South America to emerge on its westerly side in the Pacific Ocean just off the coast of what is now Chile.

A historical note here: this route in the southern tip would be known as the Drake Passage. The other well known transit around the end of South America is the Strait of Magellan, so named after the Portuguese explorer Ferdinand Magellan who sailed through the strait in 1520. The Magellan fame in history is that he is credited as the first explorer to attempt to circumnavigate the globe, but unfortunately died before completing it while in the Philippines. The journey was completed though by another captain aboard Magellan's ship named the Elcano. The two routes are different with Magellan's being the closer to the coastline than what Drake had sailed.

Now sailing northward up the Pacific coast of South America toward the area of Peru, Drake found more opportunities to continue his plundering of Spanish ships and local villages. The Spaniards had for many years searched for opportunities to ravage the local villages of the indigenous people of South America for their valuable gold amassing much treasure. Legends of the mighty Aztec fortunes and the huge golden treasure of Montezuma legend circa 1519, were giant lures to attract any greedy conqueror. The Spaniards for decades had drained much wealth from the southern continent. Drake was now taking it from the Spaniards for his Queen in England.

Drake's next success against the Spanish was going to be huge when he captures two Spanish ships in 1579 near Esmeraldas, Ecuador just north of Peru on the northwest side of South America. The first ship was said to have carried 25,000 pesos made of gold, the second ship named the *Nuestra Senora de la Concepcion* also known as the *Cagafuego*, a large three masted Spanish galleon that was carrying 28 tons of silver and 80 pounds of gold! In today's market (2022) the silver is worth more than 18 million dollars, and the gold at about 30,000 dollars. To describe what these weights are in actual dimensions, a standard gold bar is 7 x 3⅝ x 1¾ inches and weighs 25 pounds. The silver, if in bars, would each weigh about 100 ounces or 6.25 pounds and measure 7.2 x 3.2 x 0.8 inches and would total about 9,000 bars! This would occupy a space for silver bars of approximately 20 rows long x 20 rows wide x 23 high or 12 feet x 6 feet x 18 inches approximately. That would only take up a small amount of space on a ship that is 102 feet long and 20 feet wide with a cargo carrying capacity of over 100 tons. The looting of the *Nuestra Senora de la Concepcion / Cagafuego* would be the largest trophy taken by Drake during this voyage. Drake would release the *Concepcion* and its crew after taking several days to unload its cargo. The *Senora de la Conception* in 1641 would be sunk at sea in the Dominican Republic after being caught in a storm.

Drake, now in fear of a Spanish Armada chasing him with only the defenses of one ship, decided he would follow the route westward as had Elcano when he circumnavigated the globe. The voyage to circumnavigate the globe was successful as Drake does sail up to the southern California coast, across the Pacific, through the Indian Ocean to the Philippines and around the eastern tip of Africa, then up its western coast returning to Plymouth, England. Drake lands in England in September of the year 1580, three years after he began the journey. Drake is now recognized as the first Englishman to successfully sail around the earth in one journey. He had returned with so much wealth captured during his trip, that half of it was given to the Queen and was said to have been more than the crown's income for an entire year. Note, there was more than one ship that was raided during this journey, plus there were a few more ports of call that were plundered and not mentioned herein. One can only imagine the total amount of treasure that was really acquired.

I find it very interesting that the Queen did issue an ominous edict after this voyage, that all of the 59 crew members returning from the trip were sworn to secrecy about the venture, or be subject to death. Plus she commanded that all of the journals and ships logs were to be destroyed. Did she fear retaliation from the Spanish in their discovering that so much had been taken from them by her privateer? Was she concerned that the English would now be seen as more adventurous as to not be fearful of circumnavigating the globe, who possibly could claim new territory before any other countries could seize that opportunity? England and France had long been in land disputes and conflicts with each other with each claiming territory that was held by the other. Maybe there is more to this attempt at secrecy about the voyage than has been known.

Drake upon his return from circumnavigating the globe does buy a large home called Buckland Abbey in the area of England known as Devon, and makes it his final home. The home is purchased from Sir Richard Grenville, who later will become one of the first Englishmen to attempt a settlement at Roanoke, Virginia in 1585. Sir Richard Grenville, Sir Walter Raleigh, Sir John Hawkins, and Francis Drake were all related! They all would be involved in much privateering, and the early colonization attempts of the New World.

Drake was a popular figure now and it had propelled him into politics, and in September 1581 just one year after his voyage, he accepted the position of Mayor of Plymouth, England. It didn't take him long though to ascend from the position of mayor, to a Parliamentary level in the Queen's service.

Francis Drake was so liked by Queen Elizabeth I, that in April of 1581 she bestowed upon him the honor of knighthood; he would officially become Sir Francis Drake. Not bad for someone who began his naval career as an outlaw slave trading pirate.

During the next couple of years, Sir Drake would serve the crown in concerns of the navy, the colonization of America, defenses against the crown's foes such as Spain, and any local issues needing attention. Drake would be involved in a couple more major events during his privateer days for the Queen of England. The next major campaign of naval activity occurred in September of 1585, it would be a mission ordered by the Queen to attack Spanish colonies located in the regions of South America.

The Anglo-Spanish War began in the year of 1585, and by order of the Queen, Drake was put in command of a fleet of 21 ships carrying more than 1500 soldiers with the mission of destroying Spanish colonies. Sailing southward from Plymouth, England in

September, he first attacked a mainland Spanish town called Vigo, and then headed west across the Atlantic attacking the colony of Santiago located in the Verde Islands, which is off the west coast of Africa. He then continued to proceed southwest across the Atlantic to the coastland of South America where he raided the locations of other Spanish presence in the port of Santo Domingo, Cartagena, and eventually (San Agustin) St. Augustine in Florida.

The English raiding armada had been successful, and upon completion of their many preemptive strikes against the Spanish, began their journey home toward England in June of 1586.

Sir Drake decided that on his way back, to first sail north along the Atlantic coast of North America toward the English colony at Roanoke. His cousin, Sir Walter Raleigh had organized and financed the initial expedition in 1585 with the purpose of establishing an English settlement on the new continent. Drake, upon his arrival in June 1586, was hoping to see a successful new colony, but instead found a group of disgruntled settlers wanting to leave with him on his return to England. Drake seeing little hope of the colonists surviving without receiving more supplies, took them with him on his return voyage to England. Roanoke Colony was a failure that had been abandoned. Sir Richard Grenville, who was to re-supply the Roanoke colony, had sailed from Plymouth, England arriving in August just missing Drake with the departing colonists. Grenville searched the now deserted colony hoping he could find the colonists, but eventually decided to leave some men behind to continue the search while he returned to England. Roanoke was to become known in history as the Lost Colony, even the men that Grenville left to continue the search were never found.

Usually the sailing route from the Americas was to follow the Gulf stream currents of the Atlantic that flowed northward up from the Caribbean just off the eastern coastline of America, and then using the familiar Trade Winds blowing from the west toward the east to cross the Atlantic toward Europe. The ocean waters of the Atlantic actually create a clockwise current flow that is known as the Atlantic drift that helps ships along a circular voyage route from continent to continent. I wonder, how many ships throughout history have sailed past Nova Scotia on their way home?

In 1587, the English became aware that Spain was going to attack the English mainland in retaliation for all of the hostile activities against Spain's economic endeavors in South America and elsewhere. The hatred between the two differing religious ideologies — one Protestant, the other Catholic — also had not helped in their relationships either. John Hawkins, along with Francis Drake, were made Rear Admirals in charge of the English fleet assigned to defend England at sea with orders to do damage to the Spanish fleet before they could attack the mainland of England. Drake and Hawkins did severely curtail any attacks when they organized a pre-emptive attack of their own, sailing to Spain where they destroyed some 37 Spanish ships anchored in the harbor. The Spaniards would now be delayed a year while they re-organized a counter attack, determined more than ever to inflict retribution for all they had lost over the years.

As the re-grouped ships of the Spanish fleet approached the English channel in 1588, Drake would use some of his old privateer tricks to maneuver away from the main fleet and capture a Spanish ship named the *Nuestra Senora del Rosario*, which just happened to be used as the treasury ship with the funds needed to pay the Spanish Army. Not indicated in my research how much of that was aboard the ship, but it is certain to have been substantial. The Spanish armada would be defeated.

It is not surprising to know that after the defeat of the Spanish armada, that in 1592 and again in 1594, two hospitals were established for elderly and sick mariners, both being funded by Sir Drake and his cousin John Hawkins. They had the notoriety, the political clout, and I would assume they had plenty of wealth to make it happen.

Drake and his cousins had all been successful privateers who eventually became involved in the service of the English crown in one capacity or another. John Hawkins began as a slave trader, yet ended up as an Admiral in the English navy. Sir Richard Grenville would be knighted by the Queen for his duties to the crown in helping to establish an English colony at Roanoke. Sir Walter Raleigh was honored by the colony of Raleigh, North Carolina being named after him for his service in many of the crown's naval ventures.

Drake and Hawkins would do their last crusade in the name of the crown in 1595 against the Spanish held ports of San Juan de Puerto Rico, and Panama. The Queen was trying to end all hostilities with Spain by again disrupting their flow of gold and silver from the West Indies. Unfortunately for Drake and Hawkins, their fleets were no match for the Spanish forts large cannons forcing both to retreat to nearby safe shores. The tropics were not a hospitable environment for those not accustomed to the hot climate; the bacteria harboring water used for drinking would cause Sir Drake to suffer dysentery causing him to die at the age of 56. Some may call it, Montezuma's Revenge!

Sir Francis Drake did it all, starting his early life as a pirate, then becoming a privateer, and later changing his lifestyle to become a lawyer and a statesman, an Admiral, and even was elevated to knighthood by Queen Elizabeth I. But his most notable accomplish in my understanding of his history, is the significant amount of treasure he reportedly had plundered during his lifetime.

Spain and England would continue to engage each other in maritime conflict until 1604 with each side finally deciding on peace.

What all of these incidents reflect, is that during Sir Francis Drake's lifetime he did manage to acquire a great fortune in plunder, whether as a pirate or as a privateer. Did he cache any of it upon some island or secretive hiding place? Probably. Did he ever get the opportunity to go to Newfoundland or Nova Scotia? Yes.

Did he command enough manpower to do the necessary laborious constructions on Oak Isalnd? Yes. Drake and his cousin Hawkins were at one time slave traders, so if manpower was needed, they wouldn't be hesitant to ignore using that labor, especially if their captives were illiterate and of foreign tongue. What they couldn't understand or communicate to others would be almost as good as "Dead men tell no tales!"

The easiest sailing route from the Caribbean was always to navigate up the Atlantic coast using the northward flowing current of the warm Gulf Stream, then to harness the easterly trade winds to cross the Atlantic back to England. That route had long been used by seafarers who would successfully transit from one continent to the other. Sightings of the shorelines of Nova Scotia and Newfoundland would be the most northern landmarks to indicate the crossing point for the seaward voyage home before entering the vast icy waters of Greenland.

Some historians believe that Sir Drake may have seized as much as 500 million dollars worth of valuables in his travels around the globe. It all was never accounted for. Could it be on some island?

Theorist and famed author Peter Amundsen, who has appeared in the Oak Island television series, shared some of his thoughts about Sir Francis Drake that indicates Drake as a reasonable Oak Island suspect. Drake had an association with Sir Francis Bacon who was one of the leaders in the secret Christian group known as the Rosicrucian Order. The Rosicrucian Movement became active in Europe around the beginning of the 17th century and was comprised of a collective group of intellects who were writers, philosophers, scientists and others, who had united to revive a spiritual and cultural esoteric secret society.

This period in 17th century European history was a time for seeking enlightenment, and the Rosicrucian held mysterious truths about the ancient past that had been intentionally shielded from the average person for long periods of time. The Rosicrucian knowledge had originally begun around the year 46 with a small group of spiritually converted followers known as Gnostics who studied the true teachings of Jesus as taught by Mark, who is one of the twelve disciples. The early Rosicrucian identified themselves by using the symbol of the Calvary cross (crucero) with a red rose at its center; the rose represented love, passion, and sacrifice, and was first known as the Rosy Cross, later to be called The Rosicrucian.

Bacon, who himself was a lawyer and writer had used his literacy talents to help in the editing of the Latin Bible into the King James English version, and as some have claimed, who is also suspected of being an author to some of Shakespeare's writings. The writings, according to Amundsen, contained coded messages about Drake having secretly ventured to Oak Island on a mission to conceal something of invaluable importance possibly on behalf of those in the Rosicrucian. Other authors and researchers have also echoed the same theory that Bacon did use some writings under the guise of Shakespeare that did use coded messages and ciphers. Bacon was an educated person who was well versed in political matters along with a sense that some things needed to be done in secrecy as much of his clandestine naval experience had been. The reason Bacon used the writings of Shakespeare to conceal his coded messages was that Bacon's status in the secret Rosicrucian Order, which at the time was considered to be a subversive political and religious group, had to remain anonymous for fear of governmental and papal repercussions. During this era, and as with some others in history, open and free speech, contrary to the current opinions and dogma of the time were dangerous, even life threatening. One example as to this theory of Bacon's deceptive use is found in the Shakespearian play of *King Lear*. Within that play are found the broken syllables that combine to read "Sir - France - is - bee - con," interpreted to imply Sir Francis Bacon as author.

Among those besides Amundsen are Daniel Ronnstam, an amateur cryptologist, who also believes in his theory that the 90 foot stone found in the Money Pit on Oak Island hides an encrypted cipher created by Bacon. That cipher has yet to be interpreted as absolute in its message. There is plausible belief about some of the other writings of Shakespeare as being authored by Bacon and are due in part to the quandary of explaining why Shakespeare had several children who were not very literate. Why would such a talented author not share his positive abilities in the education of his offspring? Did Bacon use the image of Shakespeare's sometimes erratic phrasing in his works as a convenient ruse and smokescreen to conceal his messaging? That would be a clever ploy in diverting any suspicion away from himself that could jeopardize the activities of the secret religious society known as the Rosicrucian. Bacon was among the elite of his time and could not risk exposing himself as being a leader in a Masonic society. Messaging had to be conveyed secretly to other members of the Order in a way that was not evident to the uninformed, but necessary to record an inconspicuous valuable message for history. Interestingly, there are many scholars studying the works of Shakespeare that have uncovered dozens of ciphers hidden in his works who have attributed them to Bacon.

There are several other plausible theories surrounding the assumed connections between the Rosicrucian, Bacon, and his friend Sir Francis Drake, and I almost forgot their relationship with Sir Walter Raleigh. Thomas Bokenham, a researcher into the works of Shakespeare for forty plus years, found a cipher in Sonnet #52 that he described as stating, "New Scotland Isle, the treasure is in Mahone Bay." This could imply the possibility of treasure taken by Drake, or Sir Walter Raleigh that was acquired during their many escapades that may have been stashed on Oak Island, and was known to Bacon. An interesting note here is that on at least one occasion after a treasure seeking foray to South America, Sir Walter Raleigh would return to England without any Spanish loot. To return to England without seizing treasure from any of the Spanish ships, or encampments, surely created a conversation of suspicion among the populous and the crown. Once a hero of the English cause to disrupt the spoils of the Spanish, Raleigh would eventually be executed as a traitor in an erroneous plot to overthrow England's next crown, King James. More of Bokenham's theories about Bacon are written by Karl F. Hollenbach a contributor for The Francis Bacon Society and can be found online. Karl Hollenbach was a former Master of the Rosicrucian Order which still exists today in the U.S., and has done much research into Shakespeare's works. There will be more about the Rosicrucian Order in a later chapter.

In a book written by author Mark Finnan titled *Oak Island Secrets*, Finnan's research notes that both Bacon and Drake had worked with a couple of very talented mining engineers in Europe: one Joachim Gaunse and one Thomas Bushell. Both engineers were highly notable in their respective fields. Bushell had expertise in the construction of subterranean water passages possibly such as those found on Oak Island that control hydro inflow and outflow. Mark Finnan gives a in-depth look of what events occurred, or may have occurred in his book about Oak Island. The more read, the clearer the picture becomes of the ability to organize a potential covert event that possibly could have taken place on Oak Island. I want to mention here again that some of the ships logs (trip records) of Drake's adventures were either destroyed or never accounted for! The possibility of a conspiracy to hide something of value whatever it may be surely had enough capable players who had the opportunity and resources to organize such a mission. If so, was it really precious metals like gold or silver, or other valuables looted by Drake? Or, was it something that only the Rosicrucian knew, perhaps a protected secret passed down from the past that originated from the original teachings of Jesus? Bacon was crafty, and enjoyed encrypting messages as he was known to do. So many questions with so few clues, evidently, it was intended it to be that way.

Peter Easton

One of the earliest pirate figures in and around the Nova Scotia area was Peter Easton.

Born 1570, in Lymington, England, he was raised in a family with a long history of being loyalists to the crown who also had ancestors involved in the Crusades. There seems to be little history available for his early years as his first recognition only becomes noted beginning in 1602 when he receives a commission to become a privateer for Queen Elizabeth I. There was a need to protect the English fishing fleet in the North Atlantic around Newfoundland from being attacked by the lawless fishermen of other nations. The waters around Newfoundland and Nova Scotia were open fishing grounds offering an abundant variety of fish to supply any who would sail the treacherous waters of the North Atlantic. England relied on that food supply to supplement the needs of her people. The waters around Newfoundland and Nova Scotia were not friendly to all ships, some of which were England's enemies like the Spanish, and others were just not civil. Fishing was a lucrative business for those willing to sail these icy cold waters, and some fishing those waters did not accept competition.

Easton was awarded three British warships to patrol the waters of the North Atlantic, and in 1602, he sailed to Newfoundland aboard his ship named the *Happy Adventure*. The ship was easily identified, proudly flying the flag of Saint George's Cross that was emblazoned with a red cross on a field of white. St. George was known as the patron saint of England for his time during the Holy Crusades – a strong symbol of loyalty flown by Easton.

Easton established a base for his operations at a place called Harbour Grace, located on the southeastern end of Newfoundland in a large sheltered bay named Conception. Harbor Grace had been used as a seasonal fishing port since 1517 by the French that later would officially become a settlement in 1583. In this location, it would be easy for the *Happy Adventure* crew to anchor alongside the rest of the fleet in safe waters away from the harsh north sea's weather. It was a harbor that would give them quick access to the surrounding waters. A small fort to house the men was constructed closer to the mouth of the bay that also served a dual purpose as an outpost to forewarn the anchored fleet in the harbor of any approaching enemy.

All however did not last long as Queen Elizabeth I was soon succeeded by James I, who shortly after the defeat of the Spanish Armada signed a treaty with Spain in 1604. During this period of about two years between rulers, Easton and his crew had surely plundered the local shipping in the name of the crown while operating legally as privateers. However, James I of England had now become in charge of seafaring operations ordering a cease of all hostilities toward Spain, now at peace. Easton and his crews would no longer be able to privateer in the name of the English crown. Easton had now become in command of a navy with no legal right to fund his crews by privateering. King James I had also suspended any support to assist in their existence. Easton had been disenfranchised; it was time to abandon his mission in Newfoundland. He was on his own.

Peter Easton did though have three British warships armed with many cannons plus their devoted crews who had sworn an allegiance to sail with Easton. They decided to continue their plundering, even though many knew there was a risk of now being labeled as pirates. Their first destination would be to sail to the area known as the Spanish Main in the waters of the Caribbean where the transport of gold and silver aboard Spanish ships would still be a target. The flag of Saint George would be replaced with a flag others would come to fear, an eerie cloth of pure black fluttering high atop the main mast that would signify the dreaded color of death. The black flag may possibly be the very first pirate flag of all times, and the probable forerunner of the infamous Jolly Roger.

For the next decade, Easton would continue to plunder the seas, commandeering any ship that was not sunk by his growing fleet now numbering around 10 ships. He now had a formidable armed fleet of ships supported by a swelling army of men estimated to be in the hundreds. His escapades not only have him marauding the West Indies, but sailing to the coast of North Africa to the edge of the Mediterranean attacking anything in his way, even English ships, villages, and towns. All the while amassing a huge wealth of various valuables ranging from precious metals like gold and silver to gems and other valuables. Peter Easton had become a master of the seas, a commander of men and ships. When his fleet would be seen by others, it would cause fear seeing them sailing their way with that black flag of death fluttering in the wind high atop the ships masts.

Easton would become known as the most notorious pirate sailing the Atlantic, as he would prove in 1610 when his fleet blockaded the English from leaving their western ports located on the Bristol Channel. This was a main shipping route in and out of England, the blockade could be like a revenge against James I for abandoning Easton in 1604. It is estimated that the blockade had consisted of a fleet with nearly 40 ships all under the command of Easton. The Bristol Bay area was a major trade area even for pirates who would trade their plunder with regular merchants. There is some reference in history that the actions of Easton were in support of a very powerful family by the name of Killigrew. It has even been indicated that the Killigrew family had helped in financing Easton, but also shared in his booty. That family name is associated with the area of

Cornwall, England. Even King James I had recognized the family as being important enough to even knight a Sir Robert Killigrew in 1603. Ironically Sir Robert was later imprisoned in 1615 by James, as it could have related to the blockade, but not really certain why. The English Royal fleet were not willing to engage in conflict with such a large armada that had some of their own former warships. The economic effect of lost trade by local merchants caused them to eventually send pleas for help to the crown. The blockade ended after a period of time.

Easton had maintained his base in Newfoundland during these years and returned to that area after ending the blockade in England. The plundering of the coastal areas would continue with Easton's crews still capturing ships, taking what was wanted, including prisoners to help crew the ships and perform other duties. Were those prisoners used to possibly do tunnel work on Oak Island? Did Easton have any valuables to hide? Maybe to both questions!

Easton would refuge across the bay from his original fort at Harbour Grace at a small colony known as Kelly's Island. This is where Sir Richard Whitbourne, a representative of the English crown of King James I, had been sent to establish peace and order in the area. Colonies in the new continent were usually granted by royal charter as a form of reward to those loyal to the crown. Easton actually took Sir Whitbourne captive as a friendly hostage trying to persuade him to assist him in gaining control over the entire area. Not willing to take such a drastic stance against the plans of the crown, Whitbourne did agree however to sail to England to petition the King to pardon Easton from all charges of piracy. That release would permit Easton to return to England as a free man to live out his life while enjoying his accumulated wealth.

Temporarily with no immediate royal reprieve in hand, the pirating resumed with vigor.

It is noted that Easton with his fleet of loyal ships, would in just one expedition plunder as many as thirty ships in the waters around Newfoundland. He would capture their crews, said to be around 1500 men, forcing them into service to perform certain duties of the fleet. Easton would relocate farther to the southern tip of Newfoundland to the area known as Ferryland, fearing that Harbour Grace could be blockaded by the King's fleet should they come to retaliate against Easton. Here he would maintain a base for two years while awaiting word of a pardon. Easton had sent word to London that he would continue his ways of harassing and plundering ships, including those of the English unless he was pardoned.

Becoming impatient for word of a pardon, Easton took aim at the Azores off the coast of Spain. Here he would encounter those Spanish ships returning from the west Indies and the southern waters of Africa and the Indian Ocean. His account of any skirmishes are sketchy, but what is known is that he was very successful sailing into Tunis, located in northern Africa, capturing several Spanish galleons loaded with treasure. It is possible that most of this treasure was taken back to his base in Newfoundland. It is believed that on his return to Newfoundland, he may have captured another Spanish treasure ship named *San Sebastian* that was forced to return with him to Newfoundland. Easton did find out upon his return to Newfoundland that he had finally received his pardon. It would not be much longer after learning of his pardon that Easton disbands his fleet, distributing some of the wealth among his loyal captains, then sails off to France.

Easton, now a free man no longer considered a pirate, establishes a home on the French Riviera to become known as the Marquis of Savoy. Here in Savoy, he was known as a very wealthy man that eventually did marry and remained in Savoy until at least the year of 1620 when history no longer continues to track him beyond his age of 50.

Some estimation of the bounty that Peter Easton may have

accumulated was thought to be around two million pounds of gold alone. At today's current market value in 2022, that gold could be worth an estimated 57 billion dollars! There were also tons of silver taken from the dozens of Spanish ships encountered around the Spanish Main area and the Azores. What other valuable cargo was plundered from the ships he encountered is unknown as there could also have been jewelry, gems, works of art, precious relics, or famous writings by the great minds of yesteryear. Sailing vessels of the era were the transporters between the continents; they carried it all!

The amount of treasure that Peter Easton had acquired can potentially be considered the most of any pirate/privateer in history! He is recognized by all as being the most successful of all pirates and privateers who commanded the largest fleet, and is regarded as the smartest!

Did Peter Easton through his years of sailing around the waters in the North Atlantic find that special island in Mahone Bay, Nova Scotia that was located just 400 miles south of Newfoundland? Surely sailing only 400 miles on the sea was but a short journey for a captain who had navigated the entire Atlantic Ocean on numerous occasions. Did Easton have enough manpower to perform the excavation of the Money Pit? He was known to command hundreds of captured sailors. Was there a need for Easton to hide any of his enormous plunder, especially if the amount proclaimed to have been seized during his travels is true? Why not! It would be smarter to hide something than to have greedy eyes see its promise for the taking. Did his crews have the opportunity plus the know how to do the very elaborate construction needed for the Money Pit, along with the adjoining flood tunnels system without being discovered? For Peter Easton during his time around the Nova Scotia area, it would seem possible! Maybe even probable! But did he do it?

Other Possibilities & Suspects

Is there any other pirate/privateer suspects we can consider? Some other names often mentioned include Henry Morgan (1635-1688), who supposedly plundered around 100 million dollars after raiding Panama City in 1671. There was the infamous Black Beard (1680-1718), who swore he had buried his money where no man could ever find it except him and Satan. Captain John Phillips, who originally began his seafaring life in Newfoundland as a fisherman, but later turned pirate in 1723, eventually plundering 34 ships. He definitely knew the area waters and where to hide something. Another was Sir William Phipps (1650-1694), who had recovered treasure from two sunken Spanish galleons worth more than one billion dollars at today's prices and may have hid it on Oak Island. Calico Jack (1682-1720) was another English pirate who roamed the southern waters but plied the east coast around the New York area.

Pirating was a means of survival for many who saw it as a way of life that was different from the humdrum of the norm. It could be adventurous, rewarding, risky, and thrilling. But also dangerous, and illegal with uncertainty of tomorrow, not to mention the hardship at sea during long voyages that sometimes would last for years. Many who became pirates already had experience as skilled seamen, either as fishermen, commercial sailing merchants, or those sailors who were part of a naval crew during periods of war between countries. Pirates were sailing all around the world, not just in the Atlantic or Indian Oceans, but wherever there were navigable waters. The Golden Age of Piracy was noted as the most active time of piracy occurring during the 1650s to the 1720s. There were more pirates then the few I have mentioned here. The number of pirate captains and privateers active during this time is estimated beyond the dozens, even to be a couple hundred or more around the world. The number of actual seamen who were aboard those ships as pirates, definitely could easily be counted in the thousands.

I have my own suspicions about a certain trio rightly called the Sea Dogs that were prominent during the 15th century that included Sir Francis Drake, Sir John Hawkins and Sir Walter Raleigh. All were Englishmen who were granted the license to become privateers by Queen Elizabeth I. When you begin to examine all of the voyages that this trio were involved in, and their counterparts, there is definitely room for suspicion of their involvement in Oak Island. We know John Hawkins was an illegal slave trader who would brutally kill slaves after they were no longer an asset. One such incident that he reportedly was accused of was when he sent a large number of chained slaves to their death aboard a burning ship while the escapees became the prey for sharks. Hawkins would no doubt use slaves to dig tunnels or pits on any island, and then make them disappear so none would be left to tell the tale.

Sir Francis Drake was raised by the Hawkins family and had accompanied him on some of those slave trading trips. I am certain his activities were not always honorable either!. Sir Walter Raleigh was a very powerful ally of many, and a very capable naval commander of the English fleets. One of the associates of this trio included Sir Richard Grenville, who was sent by Raleigh to help establish the colony at Roanoke, who was himself an admiral with a fleet of seven ships. Grenville once captured a sizable Spanish treasure ship leaving the waters of the Caribbean, but upon his return to England did not deliver all of the estimated fortune he seized.

What I find interesting about this trio of privateers is that they may have had a little more than a touch of selfish greed in some of their many dealings while under the guise of sailing for the Queen. What promotes this suspicion is the unaccountable time gaps in some of their journeys and the length to complete some of those trips. Did they purposefully not report sailing to somewhere to avoid discovery of their hiding a precious cache? Would they later return to retrieve their secret stash to avoid sharing it with the crown who gave them license to plunder? Seems to be something to consider as being very plausible. In order for a reader to fully understand the implications here of this trio, or others to be guilty of any misdeeds, one would need to go beyond what is written here and explore in depth the complete biographies of those mentioned. What is written here was only meant to present a cursory historic overview and to give logical insight that only indicates the potential involvement of those uncertain participants. Events that have occurred on Oak Island are surely mysterious, with no definitive answer as of yet to any of the many questions that seem to haunt this lasting mystery.

Is there some treasure hidden on the island that was stashed away by any other nefarious individuals than those previously described? There are many tales rumored throughout history of immense hoards of treasure that were taken, but have never been accounted for, nor discovered. Some of the supposedly sunken ships of the Golden Age of Piracy may have been intentionally scuttled after stashing their loot ashore somewhere, only to be retrieved later with the crew supposedly disappearing at sea. There may even have been a secret mission for a king or queen to hide the valuables of the crown for fear of an invasion by one of their enemies. There was plenty of fighting between the European countries, much of which was on the high seas, but also over control of the various unsettled new lands. There were even conflicts occurring between the early colonies in North America. Naval military and merchant marine supply ships transported vast sums of gold, and silver along with other precious commodities on the sea lanes all along the coastlines of the Americas, Europe, Africa, and Asia. If a ship's crew decided they no longer wanted to be at sea among the dangers of war, or the perils of the sea, mutiny surely wasn't without thought! The possible greedy intentions of some to reward themselves with the ship's bounty surely could be a motive. The fact is, that there are numerous accounts throughout history describing the possibility of enormous amounts of treasure supposedly plundered that has never been accounted for nor found. We can only identify some of the possible treasures, and those who were involved. No one knows for certain what other treasures may be hidden. Should anyone have intentions on hiding a valuable cargo, or royal treasure for safe keeping, where would be a good location? Maybe an unknown island with a deep pit perhaps?

There are other different theories as to who may have hidden treasure on Oak Island.

King Charles I, a 17th century English ruler, was a greedy person who overtaxed his subjects and who was defiant toward the British Parliament's requests for a more legitimate and transparent monarchy. It was believed that the King had amassed more wealth for his own personal treasury while ignoring the needs of the kingdom. After years of non-compliance with the Parliament, King Charles lost his throne in 1649 when he was executed. The supposed treasure of the King included hundreds of paintings by such famous artists as Rembrandt, Rubens, Da Vinci, and others. The amount of gold and silver along with all the art that was not accounted for was estimated to be worth billions of dollars. Since there is no record of the whereabouts of this possibly enormous treasure, it was suspected or should I say rumored, to have sailed to Oak Island. Just another theory in this long list of mysteries.

How many suspects could be involved in this age old Oak Island mystery? I hope most are covered herein, but who knows for sure! Wait, there is one more Oak Island suspect to mention: the Knights Templar!

It would be negligent to only focus on pirates as being the main suspects as to what occurred on Oak Island. One can not omit others that may also have sailed the same waters. There has always been the tales of the early Vikings and Norsemen in these waters, maybe even the ancient seafaring Phoenician people who taught the Vikings how to use a keel under their ships for better balance, construction integrity, and maneuverability. Could that be true as some historians have thought? But there was one unique group that may have had a religious motive that could have involved a need to come to Oak Island. That group was the legendary Knights Templar. Who were the Knights Templar and why would they come to Oak Island? To begin, the Knights Templar were a dedicated medieval Christian army on land, but were known to possess a fleet of many ships capable of sailing any sea they desired. They had been sanctioned by the Catholic church and were best remembered as being the poor warrior monks of the Holy Crusades during the years of 1095-1312. They had been organized as a quasi-military group with the mission to protect the Christians traveling to the Holy Lands to witness the Biblical sites where Jesus had lived. The Knights Templar had originally begun as a group of nine honorable men coming from their homeland in France on a mission to Jerusalem. But the number of Knights would greatly increase over the years into the thousands for the defense of Jerusalem and the surrounding areas. Eventually a suspicion would arise that during their occupation of Jerusalem they had discovered something of a great religious value that had been hidden beneath Solomon's Temple. So now the Templar legend begins to become entwined with Biblical history.

It is true that the whereabouts of the Ark of the Covenant, and the Holy Grail have been unknown for centuries. Ever since the times of the Babylonians conquering Jerusalem in 597 and again in 586 BC, those religious articles of great Christian significance cannot be located. However the Holy Crusaders would eventually come to be known as the guardians of the Grail? Why? Although no one knows for certain what the Grail actually is, there are several different beliefs as to it's identity that will be discussed later.

Thus there is the need to include information about why the Holy Grail and the Ark of the Covenant may be a part of the mystery surrounding Oak Island that eventually involves the Knights Templar. This is just a brief introduction of the Knights Templar that will be explained in more detail in later chapters.

What has been written here is to give the reader an informative synopsis about who the potential perpetrator could be that is linked to Oak Island. There is more to consider for this part is only to give the reader an awareness of those that have been rumored by others to be the suspects. The reader must decide if any of the suspects written about herein can unequivocally be named as the one who is responsible for the mystery known as Oak Island. Probably not! There are too many unknowns to give a final verdict. I did not write these chapters to purposefully elude to any of the facts necessary to come to any conclusion about any of the suspects. It is all found in greater detail in the many history books written by the many great authors that have had trouble themselves in finding answers to a lot of the same questions. It is therefore necessary to continue the investigation as to who else may be the designer of the events that have occurred on this island. So please continue to read for there is more to come that will give a deeper insight about what the true treasure is, and who are the very probable perpetrators, and why they did it! The truth yet to come is stunning!

Remember, there is a need to look at all of the pieces in a puzzle before the complete picture is seen and the answer found!

The French & English Military

The search for treasure on Oak Island has led to some very fascinating discoveries over the past couple of centuries. Some of the things that have been found by the searchers range from old digging tools, to mysterious formations of stone piles, even including a few very old human bones that have been recovered from the depths of the so named Money Pit. Millions upon millions of dollars have already been invested with more projected to be spent, along with hundreds if not thousands of laborious man-hours that have been logged in seeking to discover a fortune of unknown wealth. The sky is the limit in the efforts that have been put forth into those searches which so far have not produced the anticipated outcome. The numerous searchers have been a variety of men with very successful backgrounds who employed great determination in their endeavors toward solving the mystery of the island. So far all have had their desired success elude them after succumbing to the many unforeseen legal and personal complications, money woes, law suits, equipment failures, and even the tragic loss of life for a few.

So how can anyone else expect to be successful when all others have fallen so short in uncovering the truth about what Oak Island is and what it has to share, if anything? The solution to any puzzle is to look at all of the pieces together as a whole and not just the parts alone. It seems that the search focus occurring over the centuries has been to excavate the Money Pit in hopes of uncovering a buried treasure of enormous wealth. But what if there is no treasure of the kind that is tangible, something we can hold in our hand that can be used to purchase material things that would satisfy our wildest desires? Suppose a treasure does exist there, but it is of an entirely different form than what has been storied about for hundreds of years. Could the current belief in the story about some kind of a tangible treasure be just a hoax? Is it really the treasure proclaimed to be of an ancient hoard of precious ingots and coins made of gold and silver, along with jewels and parchments of great works in literature or of historical note? What could be so precious and valuable, that it would warrant burial in a secreted vault of mysterious yet deadly construction. We really can't be definitive because it is still unknown who did what. The mystery would be easier to unravel if we knew what entity did what and when with the only real challenge being focused on how to recover the potential treasure in the Money Pit! So we must delve even deeper into what we already know to enable us to synthesize an answer by looking at all of the possibilities. It's like the scattered puzzle pieces of a picturesque scene that must be combined to re-create a picture that the viewer has never seen! The whole cannot be known until each of the pieces have been scrutinized for possible connections and then put into their correct order of importance to clearly expose a final view.

Structural remains of well planned potential landing sites for ships have recently been discovered along the shores of Oak Island. The structures were made with the use of hundreds of small boulders and numerous old large wooden beams and posts that have been uncovered by the Lagina team and prove that there was an organized involvement capable of constructing such projects, possibly occurring hundreds of years ago! With the size of the area impacted that created the sites appearing to be the remains of piers, wharfs, docks and a roadway, the amount of materials, time, and labor needed to complete the jobs could be considered a very significant event. Ships needed mooring piers and dry docks or wharfs to unload or load supplies along with performing ship repairs, or as a landing site to organize scouting missions, or make military encampments prior to engaging any foes in the new territory. Was this the evidence of a mission to solely construct the Money Pit? Then it would be conceivable that something more than a temporary landing site would be needed to accommodate the ships and their large work crews. The site would need to be like a small shipyard with the likely appearance of a commercial dock facility with many workers and others busily tending ships at anchor. In either instance, whether it be a military purpose, a need by others to secret something very valuable, or a place to repair ships and re-supply or rest, there are remnants that provide definite evidence of something more then any usual residential activity on the island would require. So what was the purpose of these suspected piers, wharfs, docks, and roadways that have been found on the island? That answer would be a very significant breakthrough toward solving the mystery of Oak Island. There is a need to further pry into the past in hope of finding that answer alone!

In the previous chapters, there has been much explanation about the suspicion of pirates and privateers possibly being involved in the creation of the Oak Island mystery. As a reminder to suggest that pirates lacked the ability to engineer and to do the type of labor the complex system of flood tunnels, underground chambers, and a pit hundreds of feet deep into the earth would require is not to be skeptical. But in reality, to only admit that it was not in their known skill set nor their usual habits to do something so complex.

There is the suspicion though of others such as the various military commands of the French or English governments as being suspect due to their involvement during the colonization of the northeast territories. The colonization efforts by both the English and French royalties would send thousands of troops and sailors, aboard dozens of ships on a single expedition that easily could have provided all the labor, skills, and knowledge to have completed any task on Oak Island.

In May 1746, French troops acting on the order of King Louis XV embarked from their homeland to the area of Nova Scotia, with the mission of attacking the British at Annapolis Royal/Port Royal to recapture the military fortress at Louisbourg. One year earlier in 1745, an English naval attack upon that fort led to the surrender by the then French occupants. Fort Louisbourg was titled as such in recognition of King Louis XIV of France after he had sanctioned the earlier colonization attempt of the land that began in 1713. The fort was located on Cape Brenton Island in the far north eastern area of Nova Scotia that had been previously used as a fishing village by Europeans because of its large safe harbor with easy ocean access. France would transform the settlement into the new lands that would become known as Acadia, sparing no expense in developing the fort. This attempt to regain control of the new territory, Acadia would be the largest and final campaign of four such attempts by the French to once again claim control over the territories of the northeast. The expedition force was under the command of the French Admiral Jean-Baptiste Louis Frederic de La Rochefoucauld, also known as the Duc d' Anville, who by trade was a map maker. He would lead some 11,000 troops sailing aboard an estimated fleet of 64 to 66 ships. Duc d' Anville was from the French aristocratic family named Rochefoucauld whose name will come up again as referred to in the book written by Zena Halpren, when she describes a link to another possible earlier Oak Island visitor Ralph De Sudley, a Knights Templar. The voyage led by Duc d' Anville would not be an easy one as unfavorable weather would prolong the time anticipated to cross the ocean. The voyagers would also experience illness like typhus and scurvy that spread among the troops and crews causing several hundred to eventually die. Some ships even became damaged during a few serious storms that occurred early on in the journey that would cause a few to return to France while others became scattered at sea during the ocean crossing. The remaining fleet of 40 plus ships would slowly reunite at Chebucto in what is present day Halifax, Nova Scotia in September 1746, after a long, three-month journey at sea. Chebucto is just a little over 30 miles north of Oak Island. Tragedy would further overshadow the mission when the Duc d' Anville would die in late September due to questionable causes. His death would leave his command to others, who eventually became frustrated due to the lack of their organizational skills, sickness, and death, calling it quits and returning to France in late October of the same year. Their mission of attacking the English never happened! Considering the intent of the mission was to attack the English and retake the Acadia area of Nova Scotia, the venture was deemed a miserable failure. It was a costly aborted mission resulting in an estimated death toll of 40% of the original task force, plus the deaths of hundreds of indigenous Mi'kmaq who came in contact with the sick French who traveled ashore.

In thinking that there may have been an attempt to create a safe, undetected and useable beachhead to develop a pier or wharf-like structure for repairs to ships or a place to secret something in a possible Money Pit, I doubt there was enough time to undergo such an involved task. There should have been some type of journal or record book that would describe that type of event that surely would eventually become exposed! With the size and type of materials used so far that have been found on Oak Island to construct the piers and roadway from the beach area, it is evident that it was not done by the labor of just a few in a short period of time! In conclusion about the known mission of Duc d' Anville having the possibility to be involved on Oak Island seems remote due to the short time spent in the area along with the devastating results of the high mortality rate suffered during the supposed military mission. Looking at the time frame of the mission that began in May, then three months of sea travel, only to return to England in September, in my opinion left no time to accomplish what is found on Oak Island!

Another theory embraced by some is that the French Military feared the potential loss of Fort Louisbourg once again to another British retake, possibly as soon as 1758. The French had held that location since 1713. That fear may have given reason to the necessity to secret treasured valuables as far away as even to Oak Island some 300 miles to the south. The French had lost control of the fort once before to the English in 1745, but did regain possession of the fort after a treaty was granted by the British. What the French had feared however, did happen; the British reneged on their treaty after only three years and would successfully seize the fort in 1758 with a surrender by the French. The English would retain control of the fort until 1768. Three hundred miles by open water is not that long of a journey for a sailing ship which could do that distance in probably a day or two at most!

These continual back and forth conflicts over control of the Acadia area created an unstable era in the region. Surely the commander of the French forces occupying the fort at Louisburg needed security in protecting any funds on hand that were necessary to sustain operations and payments to the troops. Funds to acquire necessary construction materials, labor, and food from the locals undoubtedly required a substantial reserve. The fort itself would have been considered a prime location to house a vault that could possibly contain large amounts of gold and silver coins along with any important documents detailing any military operations, or colonization plans. Any invading force would surely try to capture the valuables contained in the purse of their enemy and would undoubtedly search the fort first. A paranoid or very wise commander could however thwart such capture by securing his funds and orders elsewhere, especially since there was a continuous threat of an impending attack. It is well within reason that any commander in charge of a large colonization or military venture would have contemplated such a threat and wisely consider a contingency plan for the safekeeping of tangible assets and documents to be secured elsewhere. A location away from the immediate area that could offer a high level of protection and secrecy could easily have been chosen, such as a site like Oak Island or one of the many other Mahone Bay locations. Louisburg which is part of the Cape Brenton Island area on the northeast coast area though is a long way to voyage in unfriendly waters and with Mahone Bay having over 300 islands, the likely pick of Oak Island as a logical depository site seems to have a high negative probability!

Either of these mentioned groups may have had the ability to provide the needed amount of manpower to perform the tasks required to excavate and construct what has been discovered on Oak Island. However, in understanding the level of engineering that was used on the island, any possible suspects who may have caused those events should only be considered with great scrutiny and pause as to their abilities. Such an attempt at any successful execution and completion of an event with such enormity that occurred on Oak Island would require a high degree of engineering knowledge, countless days, weeks or even months of untold labor, and the highest commitment to avoid detection that could lead to discovery by others whom could be the enemy. They would need the window of secret opportunity!

The Mahone Bay region began to witness its first inhabitation late in the year of 1632 at La Have, located just to the south of Oak Island about 40 kilometers by sea on the east coast of Nova Scotia. It was first settled by the French with a group of 200 men led by a French naval officer on a mission to establish the area as the land to be known as New France. There they established a settlement called Fort Sanite-Marie-de-Grace. The area was for the most part abandoned in 1636 when the inhabitants were transferred to Port Royal, which had been established earlier in 1605 on the opposite side of Nova Scotia. It was not until 1749 that Halifax to the north of Oak Island was established with the first recorded land owner of Oak Island in 1753. It was not until the Shoreman Grant of 1759 that the mainland shores around Oak Island known as Chester would become populated.

There have been a few historical periods that may have required either the French or English military having a need to create a local safe storage haven for valuables of some form or another. Evidence of the type of project that was performed on Oak Island would normally be recorded as a routine requirement by military standards. What history has revealed though, there is no log or account of any such activity ever existing, nor any valid documentation being presented anywhere to indicate any such event ever happened prior to 1795, or was ever witnessed! Otherwise the mystery would have already been solved because someone would have spilled the beans! What is known so far is that the beginnings of Oak Island's mysterious Money Pit was in all probability created long before 1749 when Halifax was settled, and the first owner of Oak Island was recorded in 1753. Some time between 1636 and 1749, human traffic was minimal except for the indigenous Mi'kmaq of the north east region. Prior to the date of 1632, of course, was 1605 when the settlement of Port Royal on the western side of Nova Scotia was first settled. This seems to indicate that any major activity on Oak Island was either prior to the 1600s or was during the era of the French and British engagements of the 1700s. If the Oak Island Money Pit was done in the 1700s it would seem some blabbermouth somewhere had witnessed at least some of that type of activity!

So, there is still a need to continue the search for an answer of who, when, and what!

The Templar Suspects

For centuries, the cloud of mysteries that have surrounded Oak Island still continue to baffle the minds of those seeking to find the truth. Of all the suspects named there is one though, that has always been viewed with a cautious realization that they could possibly be the unconfirmed link to its solution, the Knights Templar. They were the Christian warriors of the Holy Crusades. There is a need to understand why they may be a suspect.

There are a few reasons for this suspicion. First is the alleged age old belief that the Knights Templar found something of extreme importance having a great religious value during the Crusades that began in the 11th century while in the Holy Lands around Jerusalem. For a time, the Knights had even occupied the area of Solomon's Temple, once the biblical site of the Ark of the Covenant. Whatever was found in the Holy Land needed to be protected and hidden elsewhere. What that something was is not recorded anywhere in history, yet there is only the suspicion that it was an extremely sacred treasure, maybe even the Ark of the Covenant. The Knights may have also found riches in the form of gold and silver, perhaps precious gems or even ancient secrets recorded on parchment all believed to have been stashed away in secret tunnels or chambers under the Temple. Secondly, there is the fact that some Knights Templar became wealthy in the years after they had returned from the Crusades. Some of that wealth by a few was thought to be from the sale of ancient religious artifacts that were taken from the Holy Lands. When the King of France eventually declared them heretics and ordered their arrest, their massive wealth and treasures were never found. The Order of the Knights Templar would basically vanish from the heart of Europe, or so most thought. It is possible many had absconded with their tangible assets and other treasure after having been forewarned of their impending jeopardy. It was estimated by some that the number of Knights Templar that probably fled the persecution by the King of France may have been as many as several thousand, who could have divided the treasures among themselves before fleeing to far away regions. Another reason for suspicion, one of which is surrounded by much controversy, is the possible connection to the group known as the Cathar, a religious sect of Gnostic people living in southern France during the 12th century. The involvement with the Cathar has a couple of aspects that may link the Knights to them, one that stems from the accusation about the Cathar practicing different beliefs than those accepted by the Catholic Church who condemned them as being heretics. Another is the eventual betrayal by the Catholic Church in their condemnation of the Knights Templar as also being heretics! The fact that both the Knights and the Cathar were condemned by the Catholic Church as practicing heresy may have united the two as allies. It is believed that the Cathar had been protecting some ancient secret religious information, or something of extreme value that then became entrusted to the Knights. Whatever that was is believed to have been eventually transported to Nova Scotia to be hidden on Oak Island!

    The formation of the Knights Templar had arose from the necessity of protecting Christians traveling to the Holy Lands in Jerusalem. As early as the first century C.E.,(Christ Era), Christians had begun traveling to the sites of where Jesus had lived. That pilgrimage had begun after the Roman Emperor Constantine would become the first Christian Emperor of the Byzantine Empire which included Jerusalem, and declared Christianity the accepted religion of the land in the 4th century. Constantine also had the Holy Church of the Sepulcher constructed in Jerusalem to protect and honor the site of Jesus's Crucifixion that was also the location of his tomb. Once Christianity had begun to be more widespread in the region, and the church had been constructed, more Europeans would begin to visit Jerusalem. Some would remain to become part of the Jewish and Muslim populations. For centuries after Christianity was established in the Holy Land, it would continue to gain acceptance, eventually spreading throughout Europe where it quickly took root. As the numbers of the new Christian believers in Europe increased, their numbers desiring to venture to the Holy Lands would also increase. Their journey in the beginning centuries of Christianity had occurred mostly without harm other than an occasional road bandit. However, that would all change when the Muslim leaders of the regions adopted their new Islamic faith in the 7th century to become followers of its founder, Muhammad. The two religions had completely opposing beliefs that would eventually be the cause of a Holy War making the Christian city of Jerusalem ground zero for conflicts that would last for centuries.

We will jump forward from the seventh century to the beginnings of when the violence between Muslims and Christians in the Holy Lands would rise to a level far beyond any local hostilities. The Christian population inside Jerusalem, a few thousand or so, had almost been completely slaughtered following a recent invasion by Turkish Muslims in 1076. Much of the surrounding areas and some other cities occupied by Christians and Jews had also been taken captive by the Muslims of the new Islamic faith. These hostilities had been gradually increasing with intensity resulting in the occupation of the local cities to alternate their citizenry between the Muslims and the Christians. This increase in the turmoil occurring in the Holy Lands had now risen to the level of a Holy War between the two different religious beliefs. During the next coming decades, disputes over the ownership of sites around Jerusalem would cause hostilities between the two groups to escalate, with each claiming a holy heritage to their own special area. The Christians had laid claim to the Church of the Holy Sepulchre where Jesus was buried at Solomon's Temple, plus a few other Christian holy sites. The Jewish had their sites. The Muslims would claim the al-Aqsa Mosque temple mount as theirs, which was the shrine of Muhammad the founder of the Islamic faith. The three conflicting religions were all fighting over the same location with each trying to eliminate the other in the area known as the Temple Mount.

During most of its historical time, the Holy Lands form the city of Antioch on the coast of Israel to the inlands of Jerusalem, remained primarily occupied by the Jews and the Christians. The pilgrimage to the Holy Lands had now become an ever increasing event. The continued popularity of this new religion that was spreading throughout Europe had even offered those among the newly devoted populations the opportunity to seek salvation. Many wanted atonement for past sins and crimes. It was the belief by many that salvation and forgiveness could be attained through the act of a pilgrimage to the lands where Jesus had walked. With the increased interest to travel to the Holy Lands there also arose the need to gain control of the unfriendly areas around Jerusalem. It was becoming more necessary to provide protection on the travel routes being used by the pilgrims. There was even a growing concern also about the need to protect the holy sites within the city of Jerusalem. That concern included the need to ensure that those sites that were still unharmed would remain unharmed and secure. Solomon's Temple along with a few of the other historical structures in the area had already been destroyed once, causing fear that what was still remaining could also be in jeopardy of destruction! Action needed to be taken!

At first, the effort to secure the lands of Jesus would be led by the bravery of the every day European Christian Pilgrims in attempting the rescue of Jerusalem. They would first though have to determine if the journey to the Holy Lands was feasible for a large group of believers which depended upon a couple of different needs. The first consideration would be did the pilgrims have sufficient enough time to complete the travel to and from Jerusalem that would be practical. These journeys to the lands of Jesus would usually take a couple of years for many of the travelers. The distance from France to Jerusalem is estimated to be about 2000 miles. Not all of the pilgrims though would come from that area, yet some would come from even farther away places like England. One can only imagine the gravity of the decision needed to be made in deciding to travel that distance and the commitment it would require. Many pilgrims were just ordinary people who could not commit to that long of a time required for this journey; many would need to return to their homeland in a timely manner to continue their normal lives. Others may not have been as restricted by the amount of time it would take for a complete round trip pilgrimage. There were even those who would make the journey with the intentions of becoming settlers in the lands. Some wanted to remain in order to be surrounded by others whose beliefs were similar to their own. Some would want to remain who could offer their help in maintaining the existence and safety of the Christian lands. Those type of considerations had to be made first before deciding which route would be the safest. There were only two route choices during the Middle Ages: either you go by the sea route or travel across the land, over the mountains, and then across the sandy desert. Both ways had their pros and cons which was not always a clear choice. Another thing to consider was how would the pilgrims defend themselves from the thieves along the land route, or from the occasional marauding Muslim ships at sea. Depending upon the travel route, the land route through Europe could take as much as two years to complete for many who would go through the Alps or Pyrenees mountains to get to the Arabian Peninsula. Once on the Arabian Peninsula, the pilgrims would be in the land of the Muslim Turks, the Islamic enemy of the Christians.

The sea route across the Mediterranean from countries like Italy or Spain to Israel was much quicker, but pilgrims could be exposed to violent storms that could cause deadly shipwrecks. The sea route also had the threat of attacks by Muslim war ships and Egyptian pirates. Either route was a choice made on faith. The Christian pilgrims coming to the Holy Lands were well aware of the horrors others had suffered, but had no misgivings to make the journey, because to the truly faithful, it was worth the risk.

The pilgrims easiest route to travel was of course to come by sea aboard ships from the various parts of Europe. They would be sailing across the waters of the Mediterranean making landfall on the coast of the Holy Lands at the town of Jaffa near Jerusalem. There they would depart for a 35-mile, two-day trek across the land, often leaving them vulnerable to groups of bandits of Arab Muslims known as Saracens that would lay in wait along the journey. These were vicious groups of militant followers of the Islamic faith who considered the Christians as unwanted invaders of their land. The pilgrims would try to defend themselves, but were often caught off guard along the trail to be ambushed or overwhelmed and outnumbered by a more powerful group of these Saracen savages. The pilgrims were always robbed of whatever valuables they had in their possession. The entire caravan, sometimes numbering in the hundreds would be totally slaughtered—men, women, even children—leaving their bodies to be found rotting by the roadside as a warning to others. Needless to say this could not continue as the Holy Land needed to be secured for the Christians.

Those attacks on pilgrims would warrant the urgent call to regain the control of the Holy Lands, especially Jerusalem. The savage atrocities against the Christian pilgrimage had stirred much anger in the European homelands, especially in France among the many Christian monks, priests, and the everyday people of faith. Pope Urban II, even aroused the ire of one Peter the Hermit, a French preacher who became so vocal in a public speech about the atrocities in Jerusalem, that he was able to amass an army of twenty thousand, plus of the angry citizenry. That would become the People's Crusade to the Holy Lands in 1095. Unfortunately for all those souls whose mission was meant to quell any hostilities they would encounter upon arrival in the Holy Lands, would instead suffer the loss of thousands of dead at the hands of a more vicious adversary, the Turks. The would-be saviors had encountered their enemies just south of Constantinople at the edges of the Holy Land. Peter the Hermit would fortunately live to join a later crusade. The loss by the brave, ill-trained and undisciplined fighters of the People's Crusade who had attempted to free the Holy Lands of its scourge, would ignite an even greater fervor among the Christian faithful in Europe. Pope Urban II, then Catholic bishop of Rome, had prayed for a more successful attempt at freeing the Holy Lands that would offer a safer passage for the pilgrimage to Jerusalem. It would only take two months for the news of the defeat to rapidly spread throughout Europe that caused a rally cry that was heard by those trained as warriors.

These warriors, some of whom were knights known throughout the lands, wanted to organize a new, larger group to attempt a purge of the savage attackers. Some knights would be coming by sea, others by land from France, Italy, Germany, or other parts of Northern Europe who all would soon begin to assemble near the lands north of Constantinople, the gateway from Europe to the Holy Land.

This new group of warriors and knights that were forming

would be known historically as the First Crusade, many of whom were from families with noble names who had previous combat experience learned from their conflicts in Europe. The need to distinguish the title of "knight" here is important, as opposed to later that of a Knights Templar. The Middle Ages designation of "knight" described a person of noble blood or who was granted knighthood in recognition of service to a sovereign land or its Lord. History has always portrayed for us the image of a knight being about an able-bodied man who was a soldier on horseback usually wearing armor. A knight was called "Sir," and that recognition of title permitted them to have many special privileges above others. Also joining this group of warriors would include many men of faith, even monks and priests. It was not unusual to have a group of clergy or men of faith as warriors because ever since the beginning of Biblical times, Christians have had to fight for their own existence. The constant struggle between good and evil occurred daily in all of humankind, and Christians were not exempt. Yes, they believed in prayer as salvation and that life should be sacred, but there was also the duty to survive with the mission to protect the weak. The Pope had even surrounded himself with a cadre of armed protectors as he was well aware that there were evil persons who would seek to harm him. To submit one's life to harm or death in defiance of evil would justify any vengeance in the defense of righteousness.

As the number of knights and other potential warriors continued to grow, becoming more organized as they did, one of the larger forces joining the Crusade would be led by a French Duke named Godfrey of Bouillon (9/18/1060 - 7/18/1100). Some spellings are Godfroi because of pronunciations. The Duke was accompanied by three of his brothers and a cousin. One of Godfrey's brothers would eventually become the King of Jerusalem. This time the Crusade was definitely going to be a formidable fighting force that was comprised of thousands of skilled archers, swordsmen, and foot soldiers, all accompanied by a well trained calvary. There would ultimately be four different armies that would be coming from different locations in Europe. They would then combine into one larger army that would ultimately total around 35,000 fighters. All of the warriors would cross the Bosphorus Strait waterway that separated the European Continent from the area of north western Turkey, once known as Constantinople without incident in the spring of 1097.

It wouldn't be long before the native Muslim Turks would attack the crusaders entering the hilly area near the Jewish city of Nicaea. That location was about sixty miles south of Constantinople but still in the northern area of Turkey. The Turks were no match for the well trained combined forces of the army of Franks (Germans), who came overland, plus the Byzantine army (former Roman Empire lands) from south eastern Europe who approached from the west by sea. The Turkish Emperor of the area quickly surrendered to this new wave of crusaders realizing they were not going to be as easy to defeat as those who came with Peter the Hermit. The two crusader armies now continued south where they were soon confronted by their next challenge from the Turkish fighters at Dorylaeum. That conflict lasted several days, becoming a standoff between the soldiers of the two combined armies against the Turkish Calvary. It would not be until Godfrey of Bouillion with his fighting force of knights on horseback leading his well trained soldiers would best the Saracens forcing them to retreat. Now the total forces of the Crusaders had finally become united, advancing farther into the lands that led to Jerusalem. Several more smaller skirmishes occurred along the way before reaching the major city of Antioch in late October of 1097.

Six months of hard fighting had not only cost the Saracens many thousands of fighters, along with losing their grip on some territory, but the conflicts had taken a toll on the crusaders as well. Many of the knights' horses had been killed, plus the provisions for the armies had been nearly depleted. It had been weather, unlike the cold that the northern fighters were accustomed, it had been hot and arid with mouth drying winds, plus their tired and wounded were in desperate need of convalescing time. The decision was made to rest before approaching the reinforced walls of Antioch, the large city was within their sight that looked a mile wide.

Antioch, a huge city covering about 1,100 acres of land was once a major landmark of the Roman Empire that now had been controlled by the Turks for over a decade. The city was still occupied by some Christian citizens in a mixed population of Jews and Muslims all totaling about four hundred thousand. Antioch in the past had been one of the main trade route hubs for the Middle East for European commerce. It also had been one of the Roman Empire's most important cities where the Romans once held chariot races in their great arena. It was known for being the city where the Biblical apostles Peter and Paul had begun Christianity. With only a third of the crusader armies remaining, there was hesitation as to when and how to attack the fortified city. Fortunately a Christian ally living within the walls of the city offered them a gift. The Crusaders would gain entry one night when the ally would open the gates from within to let the crusaders enter, but that would not happen until June of 1098. The crusaders had waited six months to complete their attack on the city, but at last they would succeed. After the siege of Antioch, a large contingent of the Crusaders would remain within the city walls until January of 1099. The younger brother of Godfrey, Baldwin of Boulougne, had left Antioch after the siege traveling to the northeast, deep into Turkey with his army to the city of Edessa. His mission there was to overtake control from the Turks who had seized the city that once was occupied solely by Christians.

Spring had now arrived as the remainder of the armies began to unite at their final destination outside the city of Jerusalem. It was now June of 1099; it had been two hard years of intense combat, and provisions were in short supply. Some of the more weary and wounded Crusaders had already departed for their homelands. The battles had drained them; they were done! The remaining ranks of the army were now preparing to seize Jerusalem by first scaling the fortified walls of the city which would not be so easy to assault. Jerusalem had been attacked many times before by different invaders since its beginning, so to protect its inhabitants and religious assets, it had been well fortified over the years for their protection. Here was a city that was once inhabited by many Christians who lately had either been ousted or killed off by the believers of Islam, now it was assumed only to be occupied by Jews and Muslims. Only about a third of the total warriors with their supporting escorts who had started the crusader campaign had survived to remain for the final assault. To be successful there was first a need to construct ladders and catapults, but most importantly, movable siege towers. These towers would be made of wood and built to the height warriors would need to gain access over the high walls of the city. The towers would be made mobile with the use of large wheels that would enable the soldiers to re-locate them at different places along the walls. The towers would be made so that they could be adjusted to rise to the differing heights of the city walls. The prep work would take time and the materials would have to be found nearby for the construction. The conquering of Jerusalem would not occur until July of 1099 as Godfrey of Bouillon along with his brother Baldwin of Bouloque would be the first warriors to have their siege tower rolled to the city walls. They then would use ladders like a bridge to span the distance from atop their tower to the city walls and then enter the grounds of Jerusalem. As the Christian army would gain entry after the opening of the gates, the bloodshed of the enemy soon would become evident everywhere, the Holy City soon would be freed. When it ended, the leaders of the Crusaders along with their remaining warriors, knights, clergy, and other escorts, celebrated, as thanks was given to God for their victory in the Holy City.

Godfrey had returned control of Jerusalem along with much of the Holy Lands back to the Christians, as he would soon become the ruler of Jerusalem. Unfortunately, in capturing Jerusalem, he would lose many of his fighters in combat. Now many of those from the accompanying armies would return back to their homelands in Europe. It was estimated that the number of knights remaining in Jerusalem was only about 300 who were accompanied by another 2000 infantry foot soldiers. There were still hostile Saracens roaming the lands near Jerusalem and the ones that were not killed in the earlier skirmishes would occasionally regroup, attacking Christian caravans that would travel between cities. There continued to be a need to protect the Christian pilgrims still coming to Jerusalem and the other nearby cities of Antioch, Edessa, or elsewhere. Complete peace would only be a prayer.

Godfrey de Bouillon, the most prominent leader of the First Crusade, would die from sickness at the age of fifty in 1100. after ruling Jerusalem for only one year. Godfrey would be buried in one of the holiest of places known, under the Church of the Holy Sepulchre in Jerusalem. Since there were no children fathered by Godfrey, his heir to the throne would be his younger brother, Baldwin of Bouloque, then becoming King Baldwin I.

Jerusalem had become a scene of destruction. It had been a war zone, but at the same time, it was a place to awe in knowing the history it held. It was home to the Holiest of Holy artifacts. There was the Temple Mount, the al-Aqsa mosque which had been the Muslim shrine, and next to it the ruins of Solomon's Temple. There was still the church of the Holy Sepulchre, and the site where Jesus had been crucified, along with the Western Wall known as the Jewish Wailing Wall. There was still plenty for the pilgrims to admire in this Holiest of Cities that would continue to lure them into their risky journey of thousands of miles to embrace the essence of its existence. As the pilgrimage would continue, so would the occasional attacks by the Saracens, so the soldiers who had remained in Jerusalem to defend them were afforded no peace. Back in Europe and beyond, word of the great success of an organized militaristic First Crusade had given others the ambition to come to Jerusalem to take part in the revitalized effort to raise the city to its former glory, hoping to assist in whatever way feasible. New pilgrims would soon begin the venture, while others of nobility would consider their contributions.

The eventual formation of the Knights Templar would now begin through the efforts of a French knight named Hugh De Payns. Payns would come to Jerusalem in 1104 as part of a small group of knights brought by another Frenchman named Count Hugh of Champagne. These were the typical knights of noble heritage that were also Christians who had come to help with the stabilization of Jerusalem. They would stay until the year of 1108 before their return to France. Hugh de Payns would come again to the Holy Lands in 1114, and again in 1119, but this time he was accompanied by a group of eight other knights calling themselves the Poor Fellow Soldiers of Jesus Christ. These knights would take an oath to follow the lifestyle of the Cistercian Monks, a religious order that was founded in 1098 as part of the Catholic Church in France. The lifestyle would be one of poverty, celibacy, prayer, rigid rules, discipline, and a commitment to the mission of protecting the Christian pilgrims in the Holy Land.

King Baldwin I had died in 1118, making his cousin, Baldwin of le Bourg to become the second Baldwin to become King of Jerusalem. King Baldwin II had already been awarded the title of Count of Edessa after his army had seized the city in 1099, then giving it back into the hands of the Christians. It would now be King Baldwin II along with the patriarch of the church in Jerusalem that would provide the knights accompanying Hugh de Payns with a small income and shelter in return for their much needed protection. The place they received for habitation was known as the Dome of the Rock, known also as part of the ruins of Solomon's Temple. The title of the Poor Fellow Soldiers would eventually change several times finally incorporating the word "temple." The temple was added to their title because that's where they had been housed thus making it an easy way to identify who they were. Now their title would be changed to Knights Templar, the protectors from the Temple of Solomon and the house of God. The casual garb of daily wear by the Knights Templar would soon be exchanged for the un-dyed white tunic of a monk. The mission of a group of Christian knights that had started with a pledge to offer protection for the pilgrims in Jerusalem had now become a supported cause in the name of Christianity.

During the next several years, not much is written about any major battles by the Knights Templar in defending the populace in the city of Jerusalem. There would be, though, an occasional attack by a small band of Saracens along the trails the pilgrims traveled from the sea port of Jafa to other nearby cities. It was a time when the threat of a major enemy attack seemed to have ebbed with the retreat of the Muslims back into the lands of the foothills near the mountains. The small population of Christians now within the city of Jerusalem certainly felt more secure with the presence of the Templar Knights who were accompanied by their warriors. That may have helped to create an atmosphere of added promise for security, but that serenity was only an imagination in the thoughts of those who knew otherwise. So many horrific conflicts in the past had taken the lives of so many that the numbers of those lost could easily be counted in the tens of thousands. The battles that had claimed such great numbers of brave Christian combatants, who also had seen the slaughter of so many innocent men, women, and children, was definitely still a memory not easily forgotten.

Word of the Templar Knights formation would travel back to Europe reaching the ears of many nobles and knights of Christian faith along with those leaders of the Catholic church.

We will stop here and do a brief re-cap of the Knights Templar first 50 years of existence before describing the next important stage in their history. It all had begun with a small group of nobles wanting to protect Christian pilgrims journeying to the lands of Jesus in Jerusalem. They committed themselves to take the pledge of obedience in living the lifestyle of a monk, but also that of a holy warrior. They were given refuge in the ruins of Solomon's Temple. As their numbers grew, the aura of being chivalrous in supporting their ranks became the importance of the Christian faithful. Funds were donated by nobles and commoners alike, and the Catholic Church gave them special rights plus privileges to acquire endowments, enabling their wealth to accrue more then the King of France. The Knights were given a very unique holy status that could only be declared by a Pope. The symbolic identification of the emblazoned "la croix pattée," the Red Cross of St. George on a white background would now command the recognition of reverence that would give rise to a mighty cause. The Knights had proven their worth by employing a skillful military style of leadership along with great displays of successful fighting tactics in battle. To die in combat for the protection of Christianity in the name of the Lord, was to die a martyr, as many would do. They had shown their obedience to the church in the adherence to the lifestyle of a monk. In just 50 years, the Knights Templar had become the most revered and feared autonomous fighting force of its time, not to forget they had also become very wealthy.

What would be their next 50 years of achievements? There would be still be another type of mission to come, one that had to be done in secrecy!

Knights Templar &
The Ark of the Covenant

The Holy Land is a part of the Middle East located on the eastern shores of the Mediterranean Sea, where the city of Jerusalem can be found. Jerusalem has long been recognized as one of the oldest cities in the civilized world. It was established around the time of 3500 BC, and its history ever since has been a complex one with many changes affecting its existence and its people. In 1000 BC the ruling king of the region, King David, claimed the area to forever become the homeland of the Jewish people. To be brief, the Jewish population had originally lived in the region of the Canaan Valley, but because of the threat of a possible famine they would leave for the more fertile lands of Egypt. After being in Egypt for almost 400 years, where they had been made slaves, Moses would lead them back to the Canaan Valley area to once again live in Jerusalem. Jerusalem has been since that time referred to as the Kingdom of the Jewish people and remains so today. Around 40 years after King David had died, his son Solomon would become king and eventually build a great temple that he would name for himself, and in it he would house the Ark of the Covenant!

During a time in Biblical history, just months after Moses had begun to lead his people from Egypt, occurring around 1300 BC, Moses received the Ten Commandments on Mt. Sinai from God. The Commandments were the moral laws, principles, and ethics for humankind to obey. Approximately one year after Moses received the Ten Commandments from God, the Ark of the Covenant was constructed to house and protect the Ten Commandments. It is said in the Hebrew Bible that the Ark of the Covenant would eventually come to be housed in a permanent temple to be built around 950 BC To safeguard it, that would eventually be King Solomon's Temple.

What does that have to do with the Knights Templar? The Knights Templar had begun as a small group of nine men who came from France to Jerusalem initially to help in defending the area from anti-Christian hostilities about the year of 1100. Christians had been traveling to the Holy Land for centuries, but their numbers had been increasing as Christianity was becoming more widespread throughout Europe. We know that the Knights were granted housing in what was said to be the remains of King Solomon's Temple. That temple at one time was intended to be the permanent home of the Ark of the Covenant! So the Knights Templar have long been suspected, and rumored, to have possibly discovered not only the hidden Ark of the Covenant along with other religious artifacts, but maybe even the Holy Grail!

So how do the Knights Templar play a possible role in the Oak Island mystery? With the Knights Templar being involved in the Holy Lands and also known to have ships that could sail the Mediterranean and beyond, it is possible that they may have had a secondary, more important mission. That mission may have been one that was less apparent than the one that was originally meant to protect the pilgrimage of the Christians. If the Templar did recover any sacred religious items in the Holy Lands, they would surely want to secret anything away from the enemies of Christianity! Maybe the Knights just wanted to protect what they found because they were fearful of how it could be used! The Ark of the Covenant was described in the Bible as having powers beyond imaginations that could destroy enemies and cities with the most catastrophic event!

Even before the crusades began in the late 11th century and before the Biblical King David had declared the region of Jerusalem to be the homeland of the Jews, there were many small battles between the Middle Eastern Muslim population and the followers of Abraham and Moses who instituted Judaism. The Muslim people believed in Islam, and worshiped a different God than did the Jewish, who during this period in the time of Moses, called their God, Yahweh. With each group trying to occupy the same land region, each claiming their own exclusive entitlement to the lands, those desires and differences by each would culminate into bias and hate which led to horrific conflicts. The Jews originally had come from the land of Judah, lying south of modern day Jerusalem, and were one of the original 12 tribes belonging to the larger family known as Israelites. The Muslim Arabs were more to the east and north in the lands of Iran, Iraq, Syria, and Turkey. The two very different religions and cultures would continually clash as each tried to gain more dominance over the other, even to the extent of complete annihilation of entire villages and cities that would leave no citizen alive. The biblical history of the conflicts between those who believed in Islam and the followers of Abraham and Moses, occurred mostly after Moses had led his people out of Egypt around 1300 BC as written in the Bible book of Exodus. The number of people that left Egypt with Moses was written as "600,000 thousand of them, besides all the women and children." The Israelites had been enslaved for about 430 years after a famine had forced them out of Israel and into Egypt as a refuge. As the Jewish people would eventually return to inhabit their lands in Israel, they would incur many enemies attempting to prevent them from that occupation. There would be many battles even between the 12 tribes of the Israelites. Hostilities arose mostly after the Ark of the Covenant began its journey from its' place of creation on Mt. Sinai then through the desert toward the Promised Land. The Ark during this time would be housed in a temporary temple shelter called a tabernacle, an ornate portable tent type structure. The Ark is a specially constructed type of chest that was not only built to contain the Ten Commandments, but was also designed to function as a direct communication device with God that enabled the Israelite leaders to hear his spoken message. It also had such powers that were described as being beyond anything humankind could imagine, and those powers could be utilized as a weapon against the enemies of the Jews, who were God's chosen people.

What did the Ark look like? After Moses had received the Ten Commandments, there was a need for its protection, and so God gave the command to build an Ark to house them. He also gave Moses very specific instructions and plans of how to build the Ark as is described in the Bible book of Exodus. Briefly, the Ark was a chest like container that was made from the special wood of the acacia tree that is found mostly in the Sinai desert. The Ark dimensions were 3¾ feet long by 2¼ feet tall by 2¼ feet wide. After the wooden structure had been completed, it was then entirely overlaid inside and out with pure gold! There was a lid cover which was also overlaid in pure gold that had two angels made of pure gold at either end facing each other. There were golden rings attached at each of the four corners through which poles could be inserted to transport the Ark. The poles, also made of acacia wood were covered in gold for use by the high priests, or by those of the Levite tribe to transport the Ark. The Ark would always be covered from view, draped with animal skins and fine blue cloths, a color symbolic of divinity. It is estimated that the amount of gold used to cover the Ark would weigh about 200 pounds! When the Ark was not being transported it would rest upon four posts for support as it was never allowed to touch the ground. It was also never to be touched by anyone, as that would be seen as the most disrespectful unholy act ever!

A tabernacle was also requested by God. A tabernacle is a place of worship that was to be a sanctuary and shelter for the Ark. The plans for its' first design and construction was to be similar to that of a large tent measuring 45 feet long, by 15 feet wide, and 15 feet in height. The materials used for the sides were sheets of fine linens that were dyed blue, purple, and scarlet with cherubim embroidered upon them. The roof would be covered with tarpaulins made from goat hair, a good water repellent, with the entire supporting framework of the structure made from acacia wood that had been overlaid with gold. All of the materials were to be joined together with golden hooks as fasteners. There are more elaborate features and other unique details about the Ark and the Tabernacle which can be read in the book of Exodus 25 thru 30.

There are many examples in the Bible that describe the powers of the Ark, beginning with the Ark being used as a tool by the Israeli army to cross the River Jordan. In the Bible book of Joshua 4: 12-17, it describes an event when the priests who are carrying the Ark are told by Joshua to step into the water at the rivers edge where it would part the waters before them. And it did! The walls of the highly impenetrable enemy city of Jericho were brought down as the Ark was carried around its perimeter led by the mighty warrior of God named Joshua. In the northern area of Jerusalem, the enemy forces of the mighty Hittite army were completely destroyed! At the time of their defeat they had been regarded as the strongest army known! Another victory for the Israelites was their defeat of another large army at the major city of Hazor located also to the north of Jerusalem. The power of the Ark was said to have been so devastating that the Israelites had to remain more than a half mile away as the Ark's power was unleashed. The Bible doesn't exactly describe the affect of the power, but it seems to infer that at least in one instance some type of a horrific event occurred emitting tremendous heat and fire as it reduced the enemy's chariots to ashes! There is actually archeological evidence that the city of Hazor did sustain some form of destruction caused by a tremendous heat source. That study was done by archeologist Douglas Petrovich who showed that the soil beneath the city of Hazor was exposed to such an extreme above ground heat source that it ignited every combustible mass of the city (conflagration) even into the ground as deep as half a meter!

The Ark, for a time, was being kept in the portable tabernacle at the city of Mizpeh, an area found in the land called Canaan. This was the Promised Land that God had told Moses would become the home of the Jews, located in an area south of Jerusalem. During the Judah occupation in the Valley of Canaan is when some unfortunate civil disputes gave rise to armed conflicts between the tribes of the Judah (Jewish) and the other 11 tribes of the Israelites. There would also be conflicts with the nearby nomadic tribes who would muster large armies numbering in the thousands. Once these two enemies became engaged they would fight to the death of all on either side of the conflict! It is astonishing to read in the Bible there were so many of these continual fierce engagements beginning as early as the book of Exodus. Those were some very violent times in Biblical history!

One of the more notable Israelite battles was with the Philistine army, a major foe of the Israelis. This one reflects the astonishing, but exclusive powers of the Ark. It revolved around a period of time when the Israelite tribe named Levi had fallen from the grace of God by not obeying the Ten Commandments. Some of the Levi tribe had gone back to their earlier beliefs of having to offer animal sacrifices that also included performing idol worship which had made God displeased. So God sent a prophet (God's chosen ones to receive communications) with a message for Eli, the leader of the Levi tribe. God was sending a warning about the actions of the Levi's tribe members, especially those of Eli's two sons. The two sons named

Hophni and Phinehas had been committing sinful acts with the young women that would come to worship at the Tabernacle in Shiloh. The two were also taking some of the offerings given at the Tabernacle to enjoy for themselves. Eli was warned that if the evil actions of his sons and the others did not cease, Eli's family would be severely punished. The father relayed the message to his sons, but they would fail to heed the warning sent from God.

The Ark of the Covenant was now temporarily housed in the tabernacle at Shiloh, a city just 20 miles to the north of Jerusalem that was near the city of Eli, so named after the Biblical character found in the book of Samuel. The Israeli army was encamped at a location called Ebenezer to the northwest about 21 miles away. That city had come under attack by the Philistines who had traveled from a city called Aphek just two miles away to the west of Shiloh. The Israelis had suffered many casualties in the skirmish, frustrated, the remaining survivors wanted to regroup and counterattack bringing with them the mighty power of God's Ark to the fight. Eli, the leader of the Levi tribe gave permission for his two sons Hophni and Phinehas to carry the Ark from Shiloh to join with the remaining Israeli army to re-engage the Philistines.

The two sons regrouped with the soldiers of the Israeli army to begin marching toward the Philistine enemy with the Ark now in their possession. The Ark, with its' immense powers, should have given the Israelis an easy defeat of the Philistines, making the two sons of Eli heroes. The Philistine army had heard of the awesome powers of the Ark wreaking total destruction on any enemy of the Jews. Their fear made them believe the end was upon them as the Jews approached with the Ark. But as the Philistines prepared to die, they would begin to desperately fight fearing their doom had come! But, to their surprise, they began to defeat the Israelis by the thousands, capturing the Ark! An Israeli survivor of the battle returned to Shiloh where he delivered the bad news to Eli. Upon hearing about the death of his two sons who had also lost possession of the Ark, Eli collapsed, falling backwards breaking his neck and causing his death. The fate of the family had been forewarned, the powers of the Ark had been withheld on purpose by God as punishment for disobeying his command!

Now the Philistines had possession of the mighty Ark that was given as a tool from God only to be used by the Jews, his chosen children. What the Philistines thought they had gained with the capture of the Ark, thinking that they had disarmed their foe, soon became their near fate. As it is told in the Bible in the book of Samuel, the Philistines took the Ark to the city of Ashbod, where they worshipped a manmade idol called Dagon, and placed the Ark of the Covenant near the idol. Awakening the next morning, they found that their idol had fallen face down on the ground at the place of the Ark. Not thinking the presence of the Ark was the cause for the mishap, the idol of Dagon was uprighted, but the next morning was found face down again on the ground. Soon after this event the people of Ashbod would began to suffer a plague of boils on their body.

Now realizing the power of the Ark was the cause of their agony, and fearing the stories of plagues brought upon others by the Ark, the Philistines sent the Ark away. So the Ark was sent to another city, but soon the plague would become widespread there also as the people again suffered and died. The Ark was sent away again to another Philistine city, but as it approached the city the people became fearful in knowing what the other cities had suffered and began begging the leaders to reject the presecence of the Ark. However, the plague had already fallen upon them eventually causing the people to suffer severe tumors that would kill many, causing widespread grieving and mournful sorrow by all! Rats had even begun to ravage their cities. After seven long months of human pain and torment, the leaders of the Philistines realized that the powers of the Ark would not be controlled by them, it was determined to be a forbidden possession! The Philistines had come to believe that the power of the Ark of the Covenant, a divine tool from God, must be returned to its rightful owner, the Jewish people of Israel. And so it would be done!

The Philistines proceeded to return the Ark to the lands of the Israelis. The city of Beth-shemesh was the nearest border area of Israeli occupation that the Philistines could safely approach. Fearing to venture any farther than necessary they would leave the Ark by the roadside. Some men from Beth-shemesh who were working in the nearby fields began to rejoice seeing that the Philistines had left the Ark. Concerned that the Commandments within the Ark may have been stolen or destroyed, several men removed the lid of the Ark to gaze inside. It is written in the book of Samuel that the men were instantly killed, numbering seventy, the people of the land would mourn those deaths, still believing that God remained angry with them. The people of Beth-shemesh believing that God had also abandoned them, wanted the Ark hurried away from their presence also, and asked their neighbors of Kiriah-jearim, just 12 kilometers west of Jerusalem, to take the Ark. There the Ark would remain undisturbed for twenty years with no bad effects on the population!

Some of the Israelites did though return to worshiping their old false idols of Baal and Ashtaroth during those 20 years. Eventually those people were advised by their leader Samuel of the city Mizpah in the north of Jerusalem, to stop their idol worship if they wished to return to the favor of God. So the idols were destroyed and again the people would pray to the Lord. A large crowd had gathered at Mizpah, now the location of the Tabernacle with the Ark to witness Samuel giving an offering to the Lord for forgiveness. As Samuel was praying, it was discovered that the Philistines were approaching the city to do battle. As Samuel led the people in a plea to God to save them, God answered with a mighty thunderous voice coming from heaven above. Confused and fearful by what they heard, the

Philistines ran away as the Israelis gave chase, inflicting death upon the enemy as they fled. The Philistines would not invade the lands of the Israelite for many years to follow. The two groups would remain enemies with deep feelings of hostility between them as each tribe wanted the other eradicated. It would be attempted again one day, when they would engage in more violent battle.

There is one more encounter with the Philistines that does need mentioning here that is found in I Samuel 14:18-23 that describes the use of the Ark again as a weapon. This time, the Ark is in the control of Saul, who would become the first crowned leader of the Israelis in late 1100 BC The people of Israel had spoken earlier that they wanted to be like other nations that had a leader they could see and touch. Samuel had been their liaison with God after Moses had died. As described in the Bible, Samuel spoke to God about what the people wanted, so he sent Saul to be the chosen one. Saul now had command of thousands of troops as King of Israel, winning many skirmishes. Having just defeated a large force of Philistines, capturing their camp at Geba, other Philistines would amass into an even larger force for a counter attack. Unfortunately the army of Saul became afraid in seeing that the Philistines now had thousands of more warriors then were originally thought, and began to abandon their positions. Saul was now left with just 600 fighters to defend his position. The presence of the Ark was at hand, so Saul asked God for His assistance. The Bible doesn't say exactly how the Ark was used that day, but the enemy became very confused, even fighting each other, and in doing so, allowing the army of Saul the chance to regroup and become victorious once more!

There are numerous references in the Bible that describe someone conversing with God, remembering that the Ark was also designed with the purpose of being a direct communication tool with God. I am sure many will believe that these conversations did occur in the presence of the Ark of the Covenant.

The Philistines would not be the only enemies of the chosen people. There were those from the indigenous tribes called the Hittites, Arameans, Edomites, Moabites, Amalekites, Jebusites, and Ammonites who occupied nearby areas of Syria, Palestine, Jordan, and even the Canaanites in the Canaan Valley.

The Ark of the Covenant would be carried from one location to another during the next 440 years after its creation. Moses had been given the instructions for its design from God when he was on Mt. Sinai. It was always required to be sheltered in a Tabernacle, the holy place of worship for the Jewish congregation and priests. After the Prophet Samuel died, and Saul had been chosen to be King of the Israelites, the mighty warrior David would slay the feared Philistine giant named Goliath. It is an interesting Biblical story that propels David toward becoming the leader of the Israeli army following the death of King Saul, and will eventually place David on the throne as the next King of Israel. As the story of David evolves in the Bible, his reign would last for 40 years before his death. But before David dies, God would give him instructions for the creation of a permanent temple in which the Ark is to be housed. However, the temple construction would only begin after David's death, leaving it to be done by his son, the next ruler of the Jewish people, King Solomon.

The use of the Ark as a weapon, but also as a communication device with God is written in the Bible to have occurred 233 times. What is written here is only just a sample of those events. It seems evident that should the Ark of the Covenant ever be found in modern times, the religious importance alone would become a worldwide event of immense importance that would cause grand celebration. But should the Ark fall into the control of those with evil intentions, wanting to unleash its unimaginable powers, the world could suffer extreme negative consequences!

There have been numerous books other than the Bible written about the journey of the Jewish people and the advent of the Ark of the Covenant. There have also been a few cinematic presentations in the form of either a full length movie or a historical documentary. One such movie most people are familiar with is the *Raiders of the Lost Ark*, starring Harrison Ford. If by chance you are not familiar with the movie, or have forgotten its theme, it is about World War II and the German Army's desire to acquire the powers of the Ark for its' use as a weapon. The movie does show the capture of the Ark by the German Nazi High Command after it was found by the movie character Indiana Jones. Wanting to see the contents of the Ark, now surrounded by many enemy soldiers, a commander removes the lid from the Ark. What occurs next is dramatized by some cinematic computer imagery displaying the release of what appears to be an enormous flow of blinding waves of colorful energy violently erupting from inside the Ark rapidly engulfing all those observing. The energy becomes so intense that those surrounding the Ark virtually disintegrate in a blinding flash, killing all of the enemy! This scene in the movie seems to reflect on the Biblical writings of the possible awesome powers of the Ark. Although this is only a movie, in actuality, during World War II the German Army under Hitler's command, did try to find the Ark of the Covenant!

The territorial conflicts along with the many differences between the religious and cultural groups of the Middle East would continue to be in constant turmoil that would last for centuries. Even after the elaborate construction of Solomon's Temple, the Ark of the Covenant would still remain to be a target for capture or destruction! The Ark would not become safe even at the location intended to be its final sanctuary!

After King David died, leaving his son King Solomon to begin his rule over Israel around 970 BC, it would take another four to seven years before the construction of the temple would be ready for the Ark. The actual location of the Temple would be atop Mount Mariah; this was the site that King David had already arraigned years before. The location that was chosen is known as the Temple Mount in the old Jerusalem city. Today most refer to the site as the Dome of the Rock, and is the shared site with the Muslim's Al-Aqsa Mosque.

The collection of the necessary materials for construction of the temple had begun while King David was still alive. Time had come to build the temple. Now there would be a great demand for skilled labor plus hundreds of regular manual laborers.

The size and complexity of the temple can be found in the Bible in the book of 2 Chronicles 2:5. I will however briefly describe here what should be the important information needed to grasp how significant this structure was in its time. First, the measurements of the temple were 90 feet in length, 30 feet in width, and 45 feet in height, with a front porch 15 feet deep, by 30 feet in length, that was surrounded by a tall courtyard wall. The exterior walls were constructed from huge white limestone stone blocks that had been quarried from the mountain areas near Jerusalem. The size of the blocks were so large and heavy that it would take a team of many oxen to drag them to the Temple location. The stone blocks would then be shaped to fit together by the work of many stone masons. The interior would be three levels high that was open to the ceiling except for a few small rooms at the rear where the altar was to be located. That was also the location where the Ark of the Covenant would be placed. All of the interior structural framing and wall paneling of the temple was made from trees that were either cedar, fir, or algum, which had been harvested from the forests in nearby Lebanon. All of the interior floors, paneled walls, doors, and ceiling surfaces that were wooden had been completely overlaid in the finest gold found! The amount of gold used to accomplish this feat was estimated to be an amount that would weigh in the tons!

In the Bible book, 1 Chronicles 29:1-9, the amount of precious metals and jewelry given to construct the temple is staggering. King David, prior to his death bequeathed all of his enormous wealth to the project plus he encouraged other leaders who were in positions of power to also donate large sums of their wealth. It is said in the Bible that the labor used to construct the temple was like the numbers of an army. King Solomon is said to have forced as many as 70,000 ordinary laborers into indentured servitude and another 80,000 men as loggers, plus 80,000 as stone cutters, with 3,600 as foremen. These laborers had been assembled from across all of the lands of Israel. There is also an untold number of skilled craftsmen, probably doing the woodwork, plus goldsmiths, silversmiths, iron, brass and bronze workers who would forge bowls, tables, lamps, or anything else needed. There were the weavers who would make the purple, blue and crimson cloth used to drape the interior walls. There were even engravers who would create art work in the images of cherubim on the golden walls. All of this labor was not found within Jerusalem during this period in its growth as there was only an estimated six thousand residents. King Hiram of nearby Tyre, a Phoenician ruler who was a friend of King Solomon, provided much of the skilled labor and the harvesting of wood from nearby Lebanon. King Hiram will be discussed later in more detail in another chapter. It was quite an enormous task to organize the completion of the temple which as you can imagine was a most opulent sight to behold when finished with all of its shiny golden structure complimented with colorful hanging drapery. It has been estimated that at today's prices the values of just the gold and silver alone would be worth a staggering $217,000,000,000! It would be almost impossible to estimate the value of the labor performed plus the cost of all the other materials such as the wood and stone. Altogether, in today's dollars, it could probably go as high as a trillion dollars!

There is much more about the temple that is written in the Bible description. Here are just a few more to know. Basically you have a rectangular building that is 90 feet by 30 feet. Now put two pillars at its 30 foot wide front on either side of the entrance door that stands at 52½ feet tall with a 7½ foot round flaring crown atop the pillars. Those two flarings would be bridged together with a decorative ornate balcony. On the inside of the temple toward the rear is where the altar is seen with the Ark sitting atop it. Placed on either side of the Ark, were two large golden angels facing each other with their wings extended, touching tip to tip that would appear to be sheltering the Ark. The two angels were sculptures also overlaid with gold, each standing 10 cubits high, or 17 to 20 feet depending upon the interpretation of the cubit.

The entire temple would be surrounded by two walled courtyards, one inner for the gathering of the priests, and an outer courtyard for the general population of worshipers. The inner walled courtyard was described as being 500 cubits by 500 cubits. In actual feet, that is estimated to be 750 feet by 750 feet, basically a square that enclosed the temple area with the wall being about three feet in height. The outer area of course was much larger. All of these measurements given are only estimates because there are several differing opinions due to the interpretation of the cubit. There is also a discussion as to the shapes of the courtyards, as there are always differing opinions by those attempting to prove their theory, there is only speculation as to their exact dimensions. But it did exist!

The completion of the temple would take Solomon an estimated seven years to complete, finishing the work around 960 BC. Knowing the estimated value of the materials used in the temple's construction, along with its basic size, layout, and its grandeur appearance, one can hopefully understand the significance of it to the people of the era! With golden angels all around, walls and floors being adorned in gold, along with its many beautiful engravings of cherubims, one can only imagine the sense of horror felt at the enormous loss when it is eventually destroyed! As you will read, the Ark of the Covenant will be jeopardized more then once after the temple was built. That is when the mysteries begin about what happens to one of the most holy of holiest artifacts in Christian history, the Ark of the Covenant!

King Solomon also had to build a new Jerusalem, as through the years of conflicts by occupations of other people, many of the old structures had been battered or burnt down to only remain in rubble. The construction of Solomon's temple would take seven years, but Solomon's palace would take 13 more years to complete standing at 180 feet long, 90 feet wide, and 50 feet high. His palace was only one of the many structures built during his reign as there were also fortresses needed to protect the area. Some of the structures built around Jerusalem were estimated to be very large. The ones necessary to house his calvary had to be large enough to stable as many as 4000 horses and their chariots! Structures were also needed to house as many as 12,000 cavalrymen. This is an important fact to remember because when the Crusades would begin later in history, the Knights Templar would use these same stables to stall their own horses along with any necessary support gear and to actually house the Templar!

Back to the power of the Ark. No person other than a chosen priest from the tribe of Levites was permitted to carry the Ark. Therefore it had to be transported in a very special way as it contained the Ten Commandments, the holiest of gifts to be given from God to humankind at that time. No one was even allowed to touch the Ark after the Ten Commandments had been placed inside. When the Temple was finished, the Ark was to be transported to the temple. It would be carried from Kiriah-jearim to Jerusalem by the Levites on a cart pulled by oxen. As written in the book of 2 Samuel 6:1-7, one of the oxen would stumble, causing one of the Levites named Uzzah to put his hand on the Ark to steady it, instantly he was killed by the powers of God. What Uzzah had done was the unholiest of all acts by contaminating the most sacred object representing God!

Even after the Ark is placed upon its final resting place on the altar in the Temple, there is one more Biblical incident needing mention. The scene is where Solomon is in the finished temple in front of the Ark as he is praying. He has come to honor God with a prayer along with an animal sacrifice to be witnessed by all of the priests and elders of Jerusalem. In 2 Chronicles 7: 1-3 it is written that as Solomon finished praying, "the fire came down from heaven, and consumed the burnt offering and the sacrifices." The sacrifices used sheep and oxen that had been slain as an offering to God as was the tradition in those days. That scripture indicates that some of the animals used had been placed inside the temple or near where Solomon was praying inferring a nearness to the Ark! After witnessing the glory of God through prayer, all that were present would kneel in prayer with praise. Here is just one more example of the unusual power that the Ark possesses.

Some have said the power of the Ark can be explained as possibly having some form of energy either electro-magnetic, or of a type of capacitor that stores electrical energy. There are thoughts also about it possibly having plasma energy, the most abundant form of energy that is known that comes from the interior of the sun. There is now even something called cold fusion. Cold fusion is a type of Low Energy Nuclear Reaction, or LENR, a hypothetical type of nuclear reaction currently being researched by many physicists. The Ark, whatever its power may be, is unknown, and will probably never be understood by humankind, at least not any time soon. The power however, as stated in the Bible, was some form of an unimaginable super energy source that was so powerful it could destroy anything! I am sure there are other theories open for discussion that will keep the scientific and theologians' minds busy for a long time!

After the Ark had been placed in the Temple in 960 BC, there would be brief mentions in the Bible during the next 373 years of it being used from time to time by different heirs to the Jewish throne. Those incidents only stated that it would be taken from the temple for celebrations, or with the army to do battle, but did not elaborate on its effect other than the defeat of an enemy. It wasn't until the year 586 BC when the Babylonian King Nebuchadnezzar attacked Jerusalem, destroying the city and the temple, that the mystery would begin as to the existence of the Ark. The temple was destroyed, being burnt to the stone walls with its treasures of gold and silver taken to Babylon, but there is no mention in the Bible of the Ark either being taken or destroyed! There is mention of some things like golden boxes and tables being cut into pieces for easy transport back to Babylon. Could the Ark have been mistaken as an ordinary box? I doubt it because the enemy of the Israelis would have certainly revealed the capture or destruction of such a prized possession and the occurrence of that event would not have escaped writings in history! Even considering the assumption that if they did find the Ark, what kind of a reaction would the Ark have generated having been violated by those not of the chosen people? Considering the powers of the Ark, it might have been a very cataclysmic event, one that evidently was not recorded, because the Ark wasn't found!

So, what happened to the Ark? Was it hidden away as some believe in the subterranean tunnels that were found under the temple that also are known to be under the city of Jerusalem as well, or even under the palace of King Solomon? Even today in the year of 2022, there are excavations currently taking place under Jerusalem that have discovered more ancient tunnels. These tunnels could have been used for the movement of soldiers, or as escape passages for the citizens during the never ending invasions by the enemies of Jerusalem. There is also the theory that the invasion by Nebuchadnezzar was known to be forthcoming according to the group known as the Talmud, the authority on ancient Hebrew history. They claim that the Ark was either hidden deep underground in a concealed chamber or vault that is yet to be found! The Ark could also have been secreted away to another safer location. One of those locations may have been back to Mt. Sinai, where Moses had received the Ten Commandments from God.

There are other theories as to the whereabouts of the Ark. One such theory is that King Solomon had a relationship with the Queen of Ethiopia named Sheba, and that her people had taken the Ark to a place called Aksum. Another location that was suspect in Ethiopia was Abyssinia where it may have been taken there on purpose before the invasion by the Babylonians.

History still remained silent on the activity or whereabouts of the Ark, even after the site of Solomon's Temple was resurrected in 516 BC by the Israelis! It took almost 70 years after Nebuchadnezzar had laid siege to Jerusalem for the Israelis to regain control of the area. It would seem that after the Israelis reclaimed their ancient city of Jerusalem they certainly would have proclaimed the presence of the Ark had it still been there! Herod the Great, who was the ruler sent by Rome to control the Jewish population before the birth of Jesus had the temple expanded and refurbished beginning in 20 BC with work lasting until the years of 10 or 11 BC. There is no mention of that occupation claiming the discovery of the Ark neither. The temple would remain unharmed until the time of the next destruction that takes place in 70 CE with the invasion of the Roman Empire. In all, it is believed that throughout historical times, the Temple had been attacked over fifty times! After all of those many centuries had passed, there would still be no mention in the Bible of the Ark's existence, nor the tablets containing the Ten Commandments. Its' known existence would not be mentioned again in any of the writings of history following the year of 586 BC!

The Holy relic of God's creation has gone missing somewhere in history. Has it been secreted away somewhere to protect it? Someone knows the truth of where it has gone. Maybe the answer can be found secreted away on an isolated island found far away from the land from which it came! There must be something of a trail to follow, if it could only be found...

Ralph de Sudeley: Part 1

There had always been an echoed belief that there was some level of involvement on Oak Island by the once very powerful and mysterious poor fellows of the cross, known as the Knights Templar. They were heralded as the fearless warrior heroes of the 12th and 13th centuries who came from all over Europe in defense of those seeking travel to the lands of Jesus. The Holy Crusades were a time of hope for the new religion of Christianity that had begun to spread across Europe and hopefully elsewhere. The Knights Templar would not only defend the travel of the Pilgrims to the Holy Land, but become the protectors of the treasures found in the Holy Land, inheriting the role of the alleged guardians of the Holy Grail and the Ark of the Covenant! With their mission so entrenched in protecting Jerusalem and those of the Christian faith coming to the Middle East, why would the Knights Templar even come to Oak Island? What could be their purpose so far away from the Holy Lands that would have them brave the forbidding seas to venture abroad to a deserted Island in an unknown land? Is there any proof of them ever having traveled to North America? Well, in a way, yes!

In two different books, one written by the late author Zena Halpren and another by Donald Ruh, they both describe the unknown journey of a Knights Templar named Ralph de Sudeley to North America occurring some time around the year of 1178.

Halpern was an expert in Middle Eastern history who describes a good amount of research aided by the findings and accounts of Donald Ruh, who collaborated with her in writing the book. Her book is titled *The Templar Mission to Oak Island and Beyond* and is well worth the read by curious minds. Even the Oak Island treasure hunters of the brothers Lagina and their partner Craig Tester were interested enough to have her appear on their TV show *The Curse of Oak Island* during Season Four, Episode Two, which was aired by the History Channel.

Donald Ruh, in his book titled *The Scrolls of Onteora-The Cremona Document,* gives an excellent firsthand extensive account of the events involving his lifelong friend named Dr. Bill Jackson. Jackson is the person who actually uncovers the trip of Ralph de Sudeley through an accidental find that leads to this historical discovery! Ruh is a close companion who, on occasion, travels with Dr. Bill Jackson on some of his pursuits to uncover the missing links that chronicle the journey of Ralph de Sudeley that leads him to a mountain in New York! The book by Donald Ruh is an absolute must read to fully appreciate what he offers to the reader in his comprehensive in-depth proof of what he and his friend have to share!

In their books, both Zena and Ruh write about the discovery of some ancient maps and navigational devices that were found by a few of the Knights Templar during their very early years of being in Jerusalem. The items were discovered in the caves below Solomon's Temple at the very beginning of the 12th century and then later were stored for a few decades in a church located in Italy. What I summarize from their books is that those items were used by Ralph de Sudeley to travel to Nova Scotia! Ruh actually describes this ancient account that was first written in Latin that chronicles the extraordinary secret journey that had been hidden for hundreds of years!

The initial Templar discovery of the items found in the caves under Jerusalem is what begins this entire journey that actually does seem to include a visit to Oak Island!

Mr. Bill Jackson in his research discovers that the Knights Templar had discovered a secret area in the underground manmade complex of tunnels beneath the temple in Jerusalem. Those tunnels had often been used by the priests and citizens of Jerusalem to escape from invaders and to secure things of value. The items and information that were found indicated a secret journey to North America for some special reason of a religious purpose! The discovery also supposedly finds a hoard of gold and the remains of what appeared to be a severed human skull. Some crudely made maps were also found that were interpreted showing a few large land masses and islands located in a far away land to the west, believed to be that of North America. The maps had been created by unknown travelers prior to the year of their discovery in 1100! The maps indicated several land masses some of which appear to be islands. One of the islands drawn is very similar in shape to that of Oak Island! In the two books, there is much detail about the de Sudeley journey of 1178 in his search for this land and its' islands. The focus of the journey describes the search for a cave location known as Hunter Mountain, which is in New York! The story continues on to describe the journey to the cave on the mountain was believed to contain certain ancient scrolls and parchments, containing knowledge of a very revealing religious event dating back to the 1st century. The scrolls also contained knowledge about tunnels filled with water; could this be a reference to the Oak Island flood tunnels? Ralph de Sudeley had recounted his journey in secret to a Cistercian monk upon returning to Europe. The written account of the journey had then been stored secretly at an abbey known as Castrum Sepulchri located in the northwestern Italian principality now known as Seborga. That area was a Templar stronghold during that time period as it was not under the influence of the king of France nor the heavy rule of the Pope! Castrum

Sepulchri is located at the southern end of the Alps which are found in the northern part of Italy near the boarder of France to its west. The region of Seborga was a very small sovereign region beginning in 954 A.D. and had been exempt from the authoritarian rule of France and Italy. The area will be discussed further in a later chapter that pertains to a certain religious sect that is important to the final chapter of this book.

The writings of the de Sudeley trip had been secretly kept in the protected archives of the Abbey of Castrum Sepulchri titled as *A Year We Remember*. Donald Ruh's book titled *The Scrolls of Onteora-The Cremona Document* referred to Onteora as is found in other history books as referring to a place in a far away land. References to Onteora were made by Middle Eastern peoples along with others, such as the Vikings and possibly even the Knights Templar. Onteora was thought to have been an unknown land very far away from the shores of Europe and the waters of the Mediterranean. There are some references that even infer that the ancient Phoenician civilization of the Mediterranean area as having knowledge of that land and may have even visited it! I find it interesting that in her book Zena's research discovered that in the language of the North American Indians the word Onteora was commonly known.

In the book written by Donald Ruh, part of that title has "The Cremona Document" which still refers to the Ralph de Sudeley document, but as it being relocated to another church also found in Italy. This will be covered herein.

How all of this knowledge about a secret journey is uncovered are the findings of a Dr. William D. Jackson that resulted from an accidental discovery he makes in 1968 at a place referred to as Bannerman Island in New York. Dr. Jackson's find eventually comes into the possession of Zena's needed expertise as she begins a collaboration with Dr. Jackson's friend, Donald Ruh. Together they begin to realize that his discovery leads to the uncovering of an incredible secret adventure.

What is referred to as Bannerman Island sits in the confines of the Hudson River near Newburgh, New York and is approximately six and a half acres in size and more formally known as Pollepel Island. The Bannerman clan, which the island gets its notoriety from, were members of a family originally from the area of Dundee, Scotland that had come to America in 1854. The island became the site of a unusual structure that was built to replicate the Scottish castle once the home of the Bannerman family in Scotland. That new structure sits near what was once the Bannerman Island Arsenal building, at one time the family business. The construction on the island was begun in 1901 when the Bannerman family business of supplying military munitions and supplies was relocated from New York city to the island for safety reasons. The father, Francis Bannerman V had started the business and later left it to his son Francis the VI who died in 1918. Only two years after the death of Francis VI in 1920, an unfortunate explosion occurred at the munitions arsenal building that would cause the site to become abandoned and eventually was turned into an historical site belonging to the state of New York in 1967. The castle site was then opened for public tours. In 1969 a fire destroyed much of the roofing to the castle and has since been considered off limits to the general public.

How this location and Dr. Jackson become important is through his visit to Bannerman Island in 1968 with some friends on a fishing trip, one of whom happens to be Donald Ruh! There the group finds two matching decorative boxes about nine inches square appearing to be ornamental with round balls on their tops, that were stuck in the mud along the shore of the old Bannerman site. Dr. Jackson takes the boxes home to use on one of his projects as an ornamental addition atop his garden gates. At some instance, the boxes accidentally come apart and are found to contain a few items. One of the items is a small, metal, three-inch disc made of brass with some ancient inscriptions interpreted to be of a Mediterranean origin, possibly as being Theban! Also found inside the box was a piece of parchment written in Latin by a G. Benvenuto, more about that name soon.

Theban is the language used during a much earlier time in history possibly being as early as the 1st century and maybe even earlier that that. I am curious as to this finding of Theban writings here but it could be related to the alphabet, possibly being one that was used in Thebes Egypt, once the prominent capital of all of Egypt. The inhabitants of Thebe were of course called Thebans. It is a possibility as some have thought, that what is called the Theban alphabet is actually part of the Phoenician culture after the Phoenicians were forced to come to northeastern Egypt after leaving Tyre, which was a part of Lebanon (circa) 334 BC.

Part of the reasoning that the Phoenicians had input into Theban is that they were the creators of alphabets including the basis for the Greek alphabet. If there really are Theban inscriptions on the metal disc that Dr. Jackson came into possession of that have Phoenician origins, then there is a whole new insight as to the research to what could be a new connection between the ancient Phoenicians, Solomon's Temple, and the Knights Templar! More discussion about that theory later.

The metal disc that was found inside one of the ornaments was determined to be some type of an ancient navigational tool of which a part of it appeared as a map to North America! Here is something that I find intriguing is that the brass metal disc had a five sided pentagram star etched into its' top. This ancient symbol will definitely come into play later in this book. Be ready when it happens! Also found inside one of the ornamental caps were some small clay tubes that contained four-inch long pieces of parchment written with an ink message that also displayed what appeared to be the French fleur de lis symbol!

Bannerman VI had been a collector of antiquities and in 1906 had purchased the brass disc that measured about one inch thick and three-inches round along with some clay tubes from a person named J. Roche. Apparently Bannerman had hidden the artifacts in the top of his garden gate posts. A clever place to hide something where not everyone would usually consider to look. The items surely had a reason to be hidden. Dr. Jackson begins to research what he had discovered and with help from some people, who have access to special investigative resources, eventually alerts other interested persons to his query. In 1971, Jackson is contacted by a man named Gustaveste Benvenuto who comes forth to sell him info about the location of a hidden Templar secret that is related to Jackson's findings. This Gustaveste Benvenuto is believed to be the descendant of the G. Benvenuto who had written the parchment dated 1820 that was found inside the Gate caps.

The books recall the involvement of two English prisoners in 1914 named Hank Roach and Liam MacDonald, who were encouraged to escape by a representative from an unknown "Order" to execute a mission to find more Templar secrets and artifacts. The search would have them travel to a church known as San Sigismundo in Cremona, Italy. That church was built in the 15th century and is located about 100 miles east of Seborga, Italy. The San Sigismundo church had been rebuilt in 1441 by a very prominent family named Sforza who were friends with Rene de Anijou (1418-1480), and Leonardo da Vinci plus others, all of whom supposedly knew of the de Sudeley trip! The prisoners, ironically, shared their heritage with other formerly involved relatives. They had close acquaintances who knew the historical knowledge passed down through the centuries about the secret of the de Sudeley trip hidden in the church! The two men did manage to escape and were able to obtain some information that had been secreted away at that church. Using the acquired secrets, they journeyed to North America ending up at a place called Hunter Mountain. Their quest was to find some hidden ancient scrolls that had been secreted away in a cave where Ralph de Sudeley, a Knights Templar had long ago searched.

What Bannerman had originally purchased was only a part of what Gustaveste Benvenuto had that could be traced back to the escapee mission. There was a need for more scrolls and items to put all of the pieces together and make sense of the map plus the disc to complete his findings. It is assumed that the search of the church only produced part of what was necessary to use the map that had been inscribed on the navigational disc, or the information about what the scrolls contained. San Sigismundo had been used as a final repository for hiding secrets possessed by the Knights Templar that were meant to be kept away from the knowledge of the Pope and the Catholic Church! The Templar secret of the de Sudeley trip was an integral part to keep those secrets hidden in the protection of Castrum Sepulchri before eventually being moved to San Sigismundo. Bannerman had only parts of what the total story could tell about the actual travel to America taken by de Sudeley. The maps and possible navigational device that were found by the Knights while in Jerusalem had originated from under Solomon's Temple and their exposure could change history! So was it Bannerman VI who was the clandestine representative of the "Order" (Knights Templar?) who in 1914 enlisted and financed a relative of J.Roche, one of the English prisoners who was sent to obtain info about the secret de Sudeley trip? It all becomes entwined with many characters of whom some have ancestry being involved dating back many centuries in protecting the knowledge of the secreted de Sudeley trip!

What is known about the Bannerman family, besides being of Scottish ancestry, is the fact that their family tree originated from one of the larger clans in northwestern Ireland known as the MacDonald. It was one of the kings of Scotland who gave the surname of Bannerman to the family after they had bravely carried the king's banner into victorious battle with the English during the 10th & 11th centuries. During one of the more famous recounts of battles between the English and Scotts, the MacDonald clan was outnumbered at Bannockburn, Scotland in 1314, but gained an easy victory when horsemen in white tunics routed the enemy. It has always been believed that some of the Templar had secretly found refuge in Scotland after 1307. Scottish heritage is deeply engrained with the warrior Templar legacy and participation. There is some definite plausibility that the Bannerman clan had ancestors and relatives that were associated either directly or in-directly with the Templar Order. Even though the Knights Templar were supposedly banished after 1307. It is true that not all were accounted for and that their ancestors still existed to further some secret mission! Knowing that the Bannerman clan name originated from the MacDonald family, we must recall that one of the prison escapees was also named MacDonald. History reflects on some things that must be judged with open minds, and sometimes accept things that may not be written in books.

    Dr. Jackson's research eventually leads him to the Vatican library in Italy where he hopes to acquire the Templar Document of the de Sudeley trip titled *A Year To Remember*. It had been a practice for some time to transfer other documents once held at the church of San Sigismundo to the Catholic library in the Vatican city. What Dr. Jackson discovers is that a man named Gustaveste Benvenuto, who was a former records keeper at the Vatican, currently had possession of the document he needed and was considering a donation of it to the Vatican. Evidently the Benvenuto family had a history of donating historical documents to the church and had possession of the Templar Document which had been a part of the family heirloom. Remembering that the original gate caps that Bannerman IV had bought in 1906, had a piece of parchment written by G. Benvenuto dated 1820 describing the disc as a navigational tool, surely indicates the family connection. The Benvenuto family had deep important connections in history that gives credibility to their having knowledge about the secret de Sudeley trip that had probably been shared by family friends named Sforza, the family that funded the building of San Sigismundo! Dr. Jackson is able through his important connections to purchase the Templar document in 1971 before it is donated to the Vatican. He now had the recorded secret journal of Ralph de Sudeley!

At this point in the course of his search for the evidence of what the trip by de Sudeley was all about, Jackson uses the info he just purchased and begins planning his travel to Hunter Mountain. Hunter Mountain is now a part of a state owned recreational park in New York that encompasses about 320 acres of hilly forested lands with its highest elevation at 3200 feet. It is necessary to note here that the author Donald Ruh and various other friends do accompany Dr. Jackson to Hunter Mountain which is not mentioned in Zena Halpren's book.

Jackson had used all of his information and research to follow the same journey in North America as did Ralph de Sudeley that led him to Hunter Mountain. The objective of the de Sudeley journey was to find what was believed to be some ancient scrolls or writings that possibly describe the secret life of Jesus and his relationship to Mary Magdalene. Stunning!

Dr. Jackson visits Hunter Mountain in the summer of 1973 in an attempt to discover the location of the cave where the ancient secret scrolls supposedly were hidden, but is not successful. Dr. Jackson, though, does finally after some more visits, discover the location of the cave in 1977 where he actually finds some artifacts that substantiate the trip as told by De Sudeley! Donald Ruh explains these trips and what Dr. Jackson finds in the cave known as Altomara's grave.

Dr. Jackson's re-cap of the de Sudeley story becomes even more fascinating as de Sudeley describes his encounters with the people of the mountain, who speak a familiar language having an accent similar to that as spoken by the Welsh who inhabit the English Isles! The people he described also wore attire upon their clothing that was similar to the type of body sash that is normally worn over the shoulder by people from Wales. Another distinguishing feature of the people was they stood taller then him and had blonde hair and blue eyes, a very Anglo Saxon appearance. There is also the encounter with a female, who presents herself as a type of local priestess, who worships a goddess and uses familiar Jewish words when speaking!

Who were these people and did they really have possession of something taken from ancient Jerusalem that warranted risking a journey across an unknown ocean in order to secret it away from those who would destroy what they had concealed.

At the beginning of the 1st century, there was much turmoil surrounding the Christian world of Jerusalem when the Roman Empire ruled the territory. Jews were persecuted and Jesus was crucified. Any historical artifacts or evidence of any Christian religious importance needed to be protected and hidden to avoid destruction. We are still not sure of the whereabouts of some of these treasures such as the Ark of the Covenant and the Ten Commandments. Ancient Jewish religious Scrolls of events taking place around 400 BC to 318 AD were hidden in the area of the Dead Sea for safe keeping only to just be recently discovered in 1946-1956. Those same scrolls written on parchment are just now being scientifically analyzed with new technology that actually is able to read what looks like charred scripture, a blackened effect that is the result of aging that occurs over the millenniums. Hopefully, as the scientific examination continues of those scrolls, humankind will learn more about the lost history of the ancient times. These are just examples of what has been hidden in times of turmoil that have occurred many times throughout the centuries. The point here is that what we seek today, that was hidden yesterday, may still be found!

Both authors go on to describe that at sometime during the de Sudeley journey there is a visit to Oak Island. During their research of Dr. Jackson's possessions, they find some maps that were hidden inside one of his books. The maps were supposedly the ones used by de Sudeley with one of the maps showing a marked X on an island shaped like Oak Island. All that is described in these two books is beyond fascinating, and actually shows drawings of the found maps and drawings of the ancient navigational metal disc like device similar to what Bannerman had discovered. The maps are hand drawn depictions of what appear to be Oak Island and other lands that are located nearby in the area of Nova Scotia. All of the writings on the maps are written in French and indicate locations of things found on the island, some of those items have been found by the Lagina brothers! In the upper right hand corner of one of the maps is found the written name of Rochefoucauld!

Earlier in another chapter the Rochefoucauld name was mentioned as coming to Nova Scotia under the title of the Duc d Anville. There is also another very important part about the map, there is no depiction of the boulders used to create Nolan's Cross! One more interesting item on the map are the words written in French that spell the year 1347. That date could indicate a second visit to Oak Island by other Knights Templar during the period of the Bubonic plaque in Europe. Another interesting piece of note is that Dr. Jackson is accompanied on his quest to Hunter Mountain by a person named St. Clair, a name that needs to be remembered. The map can be found in Zena's book *The Templar Mission to Oak Island and Beyond*. Other maps and great photos of Dr. Jackson's finds can be found in Donald Ruh's book with many different pages displaying all of the items researched by Jackson.

Thanks to both of these authors, one of whom has first hand information, have pieced together an incredible historic event through a lot of research aided with the help from many others. I do not feel it is in my area of needed discussion to elaborate on what may have been the religious secret suggested in the scrolls de Sudeley and others later pursued. That can be read elsewhere. What is important to know of the de Sudeley mission as it relates to my book is the fact that it happened, and that it is a link to the secret of what I discovered.

For hundreds of years, the Vatican seemed to have been unaware of the great importance contained in the document known as *The Year We Remember* that was recanted by Ralph de Sudeley circa 1180 to a monk at the Castrum Selpulchri in Cremona, Italy. Only a discreet few knew the guarded secret that could lead to a possible paradox for the church.

In 1994 or soon thereafter, representatives of the Vatican would purchase some documents from Dr. Jackson leaving him with an undisclosed amount of cash and an apparent caveat to honor secrecy. Dr. Jackson would leave the United States to reside somewhere in Ireland until his death in 2000.

A group from the Vatican would later go to Hunter Mountain in search of any telltale evidence possibly overlooked by Dr. Jackson and his friends, but that search was reportedly fruitless as there was still a desire by the church to seek an answer to a certain question!

These two books combined all of this information from the notes, journals, drawings, and maps that were left by Dr. Jackson that described what he and de Sudeley had pursued. What is fascinating about these two books is the details that have been disclosed about an actual documented journey to North America made by a Knights Templar named Ralph de Sudeley using artifacts that were found at Solomon's Temple! All about this discovery not only reveals the fact of the de Sudeley journey, but also gives belief that there was knowledge of an even earlier voyage to North America possibly having occurred as early as the 1st century!

This event that was hidden from the world was not an accident, but an intentional covert mission to protect and conceal certain knowledge of extreme importance that the world is still unaware of as a complete story!

This mission had originated during ancient times, coming out of the Holy Lands for some special purpose that involved coming to North America. The reason was mentioned herein, but there is more! This is only a glimpse into what comes next. The de Sudeley journey has barely been unveiled, for it is only the beginning!

If all is true, was Ralph De Sudeley the only Knights Templar to set foot on Oak Island and beyond? There may yet be cause for others to come, but why? What will be their reason to journey to the island? Who were the ones that may have come before de Sudeley that possibly created the navigational disc he used? The journey to the answer must go on!

Knights Templar & Ralph de Sudeley: Part 2

The first 50 years of the Knights Templar saw their rise to a renowned status as the protectors of Christian influence in the Holy Lands of the Middle East. Their mission as protectors, but also as avengers, had become clearer with each passing decade. Thousands had joined their ranks, partly through the act of chivalry, but most with a belief in the need to cleanse their soul through service in the name of God. Their image had become synonymous with presenting fear, yet respect, and were seen as those whom were obediently pious.

The followers of the new Islamic religion founded by Muhammad in the lands to the east of the Mediterranean Sea hated the Christians for their opposing beliefs and were willing to purge them from the Holy Lands at any cost. The Christians, however, were just as determined to stay. Their very differing religious ideologies would cause conflicts over the occupation and control of the land. The Muslim people of Turkey, Syria, Jordan, Iraq, Iran, and south into Egypt became enraged at the influx of the ones they called "The Infidel." The Christian Holy Lands were intertwined with the Jewish Kingdom of Jerusalem, but both were surrounded by this new enemy from another religion who believed their belief was the only true belief. The Muslims had deemed that all Christians should be eradicated by the blade of the sword! So the fighting would rage on for decades with the rise and fall of almost every city in the Holy Land.

Following the first Crusades, one in 1096-1099, and the lack luster performance of the Second Crusade that ended in utter failure in 1147-1149, King Louis VII of France would not venture any future attempts.

The Knights Templar, however, would continue their expansion throughout Europe after the end of the first two crusades, gaining more and more lands while building commanderies to house the new members of their ranks. Their wealth would grow from the proceeds they collected from their many agriculture harvests and rents on properties that were bequeathed to their cause. Their coffers would grow steadily, filling with each passing season. They also had become money lenders using their funds like mini banks scattered throughout Europe and into the Holy Lands, even lending monies to kings or others of royalty. Acting as bankers they even had arranged for the Pilgrims traveling to the Holy Lands a banking service to safely deposit money with them in Europe, then later be allowed to withdraw an amount out upon arrival in the Holy Lands. This system was similar to our modern day banking system that uses a simple withdrawal note. A small fee was charged for the banking transaction.

The costs to support a Templar quasi-military operation and to keep it growing was an enormous undertaking. Their funding had to be sufficient enough to acquire necessities like food, horses, armor, clothing, weapons, and constructing new commanderies, fortresses, and yes, even some churches. So any way assets could be increased would be welcomed. It was an absolute necessity to keep up with the replacement costs of things lost to battle, and the ability to further finance the expansion of commanderies into areas wanting protection. The Knights had an obligation to continue their growth if the church ever hoped to establish a safe existence for Christians in the Holy Lands.

The Knights Templar had now been occupants and defenders of Jerusalem for decades. The main housing site for them was the Temple that King Solomon had built, including his castle grounds, which was also the area of the Sunni Muslim built Al-Aqsa Mosque. The Temple area was said to have housed the Knights plus their many horses in an enormous stable located under the Temple. There are cavernous areas of limestone found under many parts of Jerusalem that are said to contain tunnel escape routes that once were used by the citizens during enemy invasions. It has been suspected that during those times of invasion, and even prior to, that valuables had been hidden in those caverns and tunnels. The caverns were created when they were quarried for use in building the many structures around Jerusalem. Herod the Great, king of Jerusalem from 37 BC to 4 BC, had Solomon's Temple re-built using limestone that was quarried from what is called Zedekiah's Cave, a five acre cavernous area located under Jerusalem. Those underground caverns were also used as burial vaults that have been found in locations under the Church of the Holy Sepulchre and the Dome of the Rock. This is important information to understand as it leads to a suspicion that something was actually found by a Knights Templar in those underground caves or quarries. What is thought to be could be one of the most important religious finds of all times!

In this part of the Templar connection to Oak Island, I will be mentioning a couple of individuals who are suspects in possibly finding some religious artifacts of great importance while in the Holy Lands. The two just happen to be English Templar Knights. I will try not to go into any unnecessary historical detail about any activities that do not directly involve these two persons. What is important during this time is how the direct involvement of what these two do and why it could affect Oak Island's history. For any history buffs wanting details specific to this time of events circa 1177 that involve Jerusalem or the Holy Lands, there are many excellent authors to read. Some very good books have been written about the era by authors like Dan Jones, Charles Addison, Piers Paul Reed, or Frank Sanello, along with some others that are very informative and enjoyable to read.

The Knights Templar in question are two English brothers who each have a very suspicious connection to the Holy Land treasures arising from their actions during the time they were in Jerusalem, and even more so what one of them does afterwards!

One of the brothers who joined the Order, was forced to do so, an unusual way to enter the cause. The two brothers of interest were William de Tracy 1133-1189, and Ralph de Sudeley 1133-1192, both great-grandsons of England's King Henry I. That is correct: Ralph de Sudeley, the Knight who traveled to Hunter Mountain in 1178. The Tracy name was their mother's maiden name, and De Sudeley was their father's. Coming from a long line of nobility, both were knights due to their ancestry and were closely involved with the church and to the king of England. During the reign of King Henry II, occurring 1154-1189, the church and the king were in accord with each other, respecting the powers each had, but later that relationship would become unstable. Young King Henry II had become a close friend with an elder bishop named Thomas Becket of Canterbury, England. In 1162, Thomas Becket was appointed to the position of Archbishop of Canterbury by the Catholic Church after the death of his predecessor, the Archbishop Theobald. It was at the urging of King Henry II that the Pope would appoint Becket to the vacated high post. After the appointment, it seemed that the new position had made Becket a different person, one that made the younger King feel inferior. King Henry II began to have disagreements about the powers of the new Archbishop as he seemed to be usurping some of the king's influence, especially over who had the most authority when administering punishment for a crime. It seemed that the church was more forgiving of most infractions committed by the subjects of the king then what the king usually would permit. After seven years of friendship between the two, that friendship would begin to wane, ultimately leading to a falling out between the two.

One evening, the King was muttering his disapproval of something the Archbishop was doing and commented, "Will no one rid me of this turbulent priest." There were four knights near enough to take the words to heart, one of those knights was William de Tracy. The words of the king were the law in those days, and knights were obliged to obey those words as part of their duty to protect the king. Although the words were probably meant in jest, maybe like a momentary feeling of anger about a friend, the knights road off to find the Archbishop.

The four knights did find the Archbishop, unfortunately what happened next became more than just a warning, and Thomas Becket, the Archbishop of Canterbury, was slain on December 29, 1170. Hard to understand how noble knights could kill a man of the cloth, but it was not unheard of in those days for a member of the clergy to be slain. There had been many disputes between the kings and others in the clergy during the Middle Ages, with each side attempting to gain influence over the other. There seemed to be a constant power struggle between the kings over who had the ultimate power to those who ruled the church. Since the control of the kingdom had been the rule of kings long before the influence of the church, it was hard for a king to yield his authority to a priest. The knights and others, who served the king would not be hesitant about enacting a bloody resolution to a conflict which was a common remedy during the Middle Ages. It was evident by this event that some noble knights had more respect for their king than they did for the Archbishop Becket.

As the king was contemplating any possible conflicts about how his ruling could affect the relationship between the secular and the religious rulers, like a power struggle, two of the knights petitioned the Pope for forgiveness of the group. The Pope did seem to have the ultimate influence above all the kings, as the Pope was revered by the people as the highest authority, while most kings were disliked as being tyrants. The King did not want to be seen as usurping the powers of the Pope in matters pertaining to the church. If the Pope could forgive the four for the assassination of an Archbishop, then in the eyes of the people it would be hard for the King to punish them. The knights had hoped that any punishment the Pope would administer would at least be less severe than any retribution administered by the king. They were correct, in a way! Pope Alexander III did forgive all four of them! But with a condition. He ex-communicated all four of the knights who were de Tracy, Reginald Fitzure, Hugh de Morville, and Richard le Breton. On top of the excommunication, the pope also banished the four to Jerusalem for a period of 14 years requiring them to serve in the Order of the Knights Templar as penance for their sin. The King yielded to the decision by the Pope!

Having to live the life style of a monk, as the Knights Templar were compelled to assimilate, obligated one to become pious while risking their life fighting against the Islamist enemy. That was going to be the only way they could earn forgiveness that would enable them to be accepted back into the good graces of the church. Even King Henry felt he shared some of the guilt for the event as he would publicly shame himself when he wept at Becket's tomb. The church had once again exercised its influence as being superior above all other leaders, reinforcing the obedient status it required of the Templar Knights to answer to only one authority: the Pope!

The eventual outcome of this incident would have a lasting impact on the future of the Knights Templar throughout Europe.

Ralph de Sudeley, who by some accounts was already a Knights Templar in England due to his noble heritage, would travel to Jerusalem after this event to join his brother, William de Tracy in 1182. Sudeley had been awarded the title of command Knight at Petra, where he would occupy the fortress at Jabal el Habis that overlooked a valley in the southern part of the Dead Sea. Ralph de Sudeley had returned to England in 1180 from what was known as his voyage to Nova Scotia in 1178 as was written about in the previous chapter. This is only a theory to believe that he was on another mission for either the church, or the Templar Order! This area of his assignment was a major trade route from Gaza in the west, and to the north toward Damascus. Petra was also known as the location of the mines of King Solomon where much copper had been mined during his reign. Copper was important during the construction of Solomon's Temple, it was used to create bronze lavers which are the large water containers used by the priests to cleanse their hands and feet before entering the Temple. The requirement of the lavers were included in the instructions for the construction of the Temple that Moses had received from God. The copper was also used to make everything from cooking utensils, jewelry, lamps, weapons, art works, or anything that was useful. All of the things that copper could be used to make also made it a valuable trade item, but copper may not have been what De Sudeley was seeking. In the area of Petra is the mountain named Jebel al-Madhbah, when translated to English means "mountain of the altar." It was thought by historians that some form of a religious treasure had been found at that location possibly during the time of the Knights Templar occupation in that area. What was thought to be found was some type of a golden chest, or a chest that contained gold and golden items!

It is not very well recorded as to the travels of de Sudeley during this time, but it is suspected that between his brother and himself, something was found while Ralph was in the Holy Lands. Why that assumption? When Ralph de Sudeley returns to England in 1189, he begins to purchase lands where he will construct a large manor house, or estate home, titled Herdwyke, which eventually encompasses about nine acres in the county of Warwickshire. He also would pay for the construction of what is known as a preceptory, a compound that provided housing with training facilities solely for the use by the Knights Templar. de Sudeley had come from a noble family, but they were not believed to be financially able to subsidize what he was doing. Remember, a Knights Templar had to give up all of their worldly wealth prior to joining the Order and were forbidden to personally profit from their service. So, what happened? With all of the tunnels under Jerusalem plus all of the caves that were being excavated at Petra, it is very possible, or even probable that if someone was really searching with a purpose, they just might get lucky enough to find something! It is well known and documented that during King Solomon's reign, he collected billions in all forms of gold and silver as payments along with receiving many types of gifts for his realm. There is also a fact that all of the holy relics of the past have never really been accounted for. We do know that one of the artifacts supposedly controlled by the Knights Templar in Jerusalem was a piece of wood from the original crucification cross of Jesus, it was usually carried into battle by the Knights Templar. "The Cross" was found in Constantinople in the 4th century where it had been held in the Church of the Holy Sepulchre until it was captured by the Islamist Saladin of Egypt in a battle. The Cross was offered in a truce agreement in 1219 by another Arab Egyptian Sultan, the Ayyubid al Kamil, after declaring an eight year truce with the Knights Templar. But according to some historians, the cross was never delivered. Others say it was. There had been several invasions of Jerusalem throughout history that had destroyed the Temple several times with the invaders pillaging what valuables they wanted as trophy. But none actually claimed that those treasures were any of the ones most revered by Christians! It could be that any historical record of something like the Ark of the Covenant, or the Holy Grail would surely have been touted by any invader as being the capture of the ultimate prize! Since there has never been any claim of certain artifacts being taken or found, then they could still be somewhere yet to be discovered! That does not rule out the possibility that there were still undiscovered treasures remaining elsewhere in a very secretive, but secure place.

There are some accounts of crusader members along with a few of the Knights Templar returning to their homelands in possession of holy relics of all sorts that supposedly had been taken from the Holy Lands. Whether all of those accounts were true or not, today there can be some relics found displayed in museums and in a few churches! Any religious artifact from those areas was a very valuable commodity which could be sold for an extremely high reward. It is recorded that de Sudeley did list some of his items in an asset report titled "Feet of Fines" as being "old relics." Those items were not particularly identified, so something could have been of a religious significance. The Knights Templar had long been associated with the belief that they were the protectors of the Holy Grail! It was never displayed by them though, although most historians believe that the Grail is a chalice that was used by Jesus at the last supper! There are differing opinions though as to actually what is the Grail. We will discus this issue later as being a possible connection to Oak Island!

The most important issue about Sir Ralph de Sudeley, a Knights Templar, is that upon his return to England he did things only the very wealthy could do. Was there a hoard of gold and silver retrieved from some hidden stash that was secreted away during the days of Solomon's Temple that others had previously missed? Did his brother, de Tracy, find treasure while exiled in Jerusalem during his time while occupying the remains of Solomon's Temple and al-Aqsa mosque? Did they find something of great value in Petra, possibly at the Mountain of the Altar? Did the brothers conspire to have de Sudeley smuggle something back to England, or elsewhere? Sadly, Ralph de Sudeley's brother, William de Tracy died in 1189, about the same year that Ralph returned to England. Another important indication of the immense wealth that Ralph de Sudeley had acquired, was when he funded the Third Crusade in 1189 that was led by King Richard the Lionheart of England. Ralph de Sudeley would die in 1192 fighting in the Holy Lands, leaving the Herdewyke preceptory to the Order of the Temple Knights.

To this day, there is a strong suspicion that Ralph de Sudeley did posses some form of treasure that he, or his brother may have discovered in the Holy Lands that was used for their own benefit. There are those who do believe there really was a treasure of great religious significance discovered in the Holy Lands. That belief about treasure has lured some in the past to even search the grounds of the Herdewyke estate in attempts to find traces of such, but to no avail. If de Sudeley did take precious artifacts from the Holy Lands, did he sell them? Otherwise, how did he have so much wealth? Who possibly could be the buyer offering enough funds to finance the expensive building cost of a castle like the estate named Herdewyke? It had to be a pretty substantial windfall for de Sudeley to also be able to fully support a Templar preceptory, along with the ability to provide enough funds to subsidize a Third Crusade! The kings of various countries were already borrowing money from the Templar banking system, so the local kings can be ruled out as being non-participatory. The Catholic Church? Maybe. But why would the church subsidize that amount of funding necessary to do all of what de Sudeley was doing? What could be so valuable to them that it would justify such a huge financial transaction to fund these commitments? Assuming such an event did occur, the price surely paid was an enormous amount that was given by a surreptitious buyer. There is no record of any such transaction! If the funding necessary to pay for all of the buildings construction was exchanged for gold or silver, maybe even precious gems, were they taken from the Holy Lands? This movement of such a large sum of finances had to be concealed, maybe even under the discrete direction of the Knights Templar Grand Master? Could there have been such an exceptional historic, or religious find in Jerusalem that its existence needed to be protected in absolute secrecy until the dynamics of history were ready for it to be revealed to the world? There could be a couple of answers to this suspicious event. First, was there something that the church had discovered that was part of Ralph de Sudeley's find they did not want to share with the world? Or was it the treasury of the Knights Templar funding de Sudeley to keep something away from the church! There was but one group responsible for the protection of any religious historical artifacts: the Knights Templar! The selling of a highly prized, or an extremely valuable and very sacred historical religious item could easily explain a reason for the sudden wealth of Ralph de Sudeley. It could also be reasoned that a part of the deal required de Sudeley to build another Knights Templar preceptory, maybe to guard his find, or was it as an atonement for violating a sacred trust!

In reflecting on the history of Ralph de Sudeley's travels, which are now known thanks to the uncovering of the document *A Year We Remember* as explained in the previous chapter, some thought is given as to who was this person? In recalling the travels of de Sudeley, there is his recorded journey to Oak Island and beyond in 1178 returning to England in 1179 or 80. Then he leaves England in 1182 to travel to the Holy Lands under orders from the Knights Grand Master to establish a new fort at Petra and to visit his brother. Ralph de Sudeley returns to England in 1189 and funds the building of Herdewyke plus subsidizes the next crusade for Richard the Lion Heart! It almost seems that de Sudeley was a special knight who traveled on certain missions for either the church or on the orders of the Templar Grand Master. The dates of his travels coincide with what is now known about him, and the recorded activities of his adventures give rise to the suspicion he did not act on his own authority. This is only my thought. You be the judge!

One final thought, if Ralph de Sudeley, or his brother, William de Tracy, did acquire some things from the Holy Lands of such an evident value. What are they? It has been said throughout history that the Knights Templar became the guardians of the Ark of the Covenant and the Holy Grail along with other religious items that are considered to be the most sacred artifacts of all time! Are those items really what they are thought to be? Do they still exist? Is there any real hope of ever finding them? Whatever they are and wherever they are is still a huge question! Will the future give us an answer? Possibly!

So What is the Answer?

There is no easy explanation as to how this mystery has become the ultimate challenge in understanding what has happened on that odd-shaped island located in Mahone Bay, Nova Scotia. By now, I think we have sufficiently examined most of the potential suspects that others have researched who possibly could have been involved in what created the Oak Island mystery. Each of those suspects' potential to commit such a feat, and the reasons to do so have been scrutinized very carefully. Yet those that have been discussed so far have not yielded any concrete evidence as to their absolute involvement, but are only left to the speculation as being suspicious. What needs to be examined now is to look at what else there may be other than the named suspects that could offer any clue as to the mystery. The elusive answer that can unravel the mystery must be found by using a different approach!

How I discovered the real secret of Oak Island was entirely by chance resulting from a determined process of elimination. Just as you have read about all of the accounts by the many potential suspects that may have been involved in creating Oak Island's mystery you begin to realize that these are probably not the actual participants. With all of the suspects most others have considered over the years being basically eliminated by my analysis, who else could it be, and what was their true purpose in choosing Oak Island, something is still escaping reason. There had to be a further examination beyond what ordinary processes have revealed to date. Maybe the answers we are looking for don't even entail any kind of a tangible treasure, and the who, what, and when, is actually something more, like possibly the answer to a different ancient mystery! Could it be the reason why we can't definitively name any one of the suspects as being the true originator of the Oak Island mystery is because we have been searching for the wrong answers, the wrong participants, and for the wrong reason! Could it be that the treasure is not gold or silver, or other valuable artifacts, but instead as some have suggested, the answer to finding something worth more than all of our earthly riches!

I began to realize that the only way to decipher what has happened to make Oak Island what it is was to dig deeper into the library of research I had created through the years. All of those dog-eared pages in books written by others who had opined about Oak Island along with the many hours of viewing articles that I had researched on the Internet had plenty of information. The TV series aired by the History Channel since 2014 had showcased many good details about their findings that kept audiences and myself interested for years. Out of necessity, I had amassed a small mountain of notes kept on 3x5 cards, Post-Its, and several spiral ring note books filled with pages of my writings with the more important facts highlighted in yellow marker or red ink. There were too many things to remember, too many little details to not forget. A lot of the material out there in book form, or written in articles, echoes much the same about a possible treasure, or who may have buried something. All of those resources out there offer plenty of great reading by a lot of talented authors who have done their due diligence in trying to present the best picture possible of the story being described. I thank all of them for what each has contributed in their efforts in the search for the solution to the mystery. Every detail that can be gleaned from their knowledge, either minute or major, can contribute toward what is needed to complete an investigation. It's all of the little pieces that when joined make a puzzle complete. The more you read the better the chances are of learning something new. I found that each of the books or articles researched did reveal a little something additional in their differing knowledge that added to the view of the whole picture. You just had to pay close attention to what was there.

A couple of the books that I had read heightened my interest a little more than some of the others. Those two books, one written by Zena Halpern and another by Donald Ruh, are definitely must reads! In both of these books they each describe the amazing secret journey to Oak Island in the year of 1178 by a Knights Templar named Ralph de Sudeley. de Sudeley had been one of the Knights who had spent time in Jerusalem near the sites of Solomon's Temple. The combined investigative research by these authors ,which took over a decade to compile, was fascinating with their findings very relevant in historical importance! There have been other books written that have shed light on other Oak Island suspects, but especially the one by Donald Ruh offers a full accounting using firsthand knowledge about events that are validated with plenty of documentation! This exposed knowledge about such an event not only gives credence that de Sudeley, a Knights Templar, had found actual knowledge of how to navigate the sea from Europe to North America, even before Columbus, but that the journey had been some kind of a secret mission that was initiated because of an ancient find once found in the Holy Lands!

In the book titled *The Templars and the Ark of the Covenant*, author Graham Phillips also writes about some kind of treasure that de Sudeley supposedly did discover while in the Holy Lands. The treasure could possibly have been the Ark of the Covenant or something similar of a religious importance. To briefly re-cap the de Sudeley episode, after he returned to England around 1189, he had become wealthy even acquiring a large estate estimated to be about nine square miles in size known as Herdwyke. There he built a large main house, and later, even adding a Templar preceptory similar to a military base capable of maintaining an estimated thousand monk like Templars. His instant wealth could have come from either found gold or silver, or his compensation from the sale of some sacred religious artifacts he had found in the Holy Lands. Interestingly, the Knights Templar order would become the eventual heirs to the Herdwyke preceptory after de Sudeley's death.

It was believed by some that there was some kind of treasure taken from the Holy Lands that de Sudeley had hidden in the walls of the preceptory. One such person who believed in that rumored assumption was none other than the Englishman, Sir Walter Raleigh (c.1552-1618) who searched the premises leaving no answer as to any discovery. I would think everything would have been gone by the time when he looked or was a treasure that was supposedly hidden within the walls just another rumored hoax. Although earlier, I had written in another chapter that Sir Francis Bacon supposedly had encoded a message in one of Shakespeare's writings stating that there was treasure from Raleigh that had been hidden on Oak Island. Bacon didn't describe what the treasure may have been. Was there really something found at Herdwyke that is linked to Raleigh's treasure? Whatever that may be! If there was any treasure found that had been part of a Holy Lands discovery by de Sudeley, I am sure Raleigh would not have admitted to finding it! But if that did happen, why would Raleigh take it to Oak Island? An interesting string of thoughts to ponder.

Another good read is the book *The Knights Templar in the New World, How Henry Sinclair Brought the Grail to Acadia* written by William E. Mann. In his book, Mann presents an interesting theory about Sinclair in his journey to North America that can only be fully appreciated by reading the whole story he has written. Mann even refers to some of the findings by Zena Halpern. I will make reference to some of those findings by Mann in my writings. There are other books that have been around for years that include some Templar events that also give some inference about the possibilities of the Templar coming to North America. Some other books are more focused on details about the battles of the Templar, or their organizational accomplishments and can become too involved unless you are an historian interested in the chronology of their day-to-day history. That is fine for those that like to have the in-depth critique. It is not necessary, though, for this book!

So what does all of the research about pirates, Templar Knights, and all of those who have searched the island have to offer as an answer of who, what, when, or why. Zilch! Those questions are still nothing but just empty blanks left for answers!

There is one more area to be covered. What about any of the items found on the island that could potentially be a clue that can lead to an answer? Over the decades of searching the island there have been many artifacts found that could tell what has happened there. What evidence there is has been slow in forthcoming to those trying to solve the mystery! A piece here this decade, another piece 40 years later; it has been very difficult to put the pieces together as most seem not to be connected enough to form an absolute conclusion that could reveal a solid answer!

It is time to look at what is physically there to see if any of those things can unlock the mystery!

Symbolic Clues Leading to the Answer!

We have previously examined most of the items found on Oak Island and what their possible intent was for being placed there. The question still is open though as to whom may have put them there at each location. There seems to still be plenty of things left to consider. What has not been recognized is the idea that something on the island is perhaps worth examining more closely for the possibility of it containing a valuable clue that has just been overlooked. Is there any kind of symbol, shape, structure, or carving on the island that we should examine more closely that could possibly be more significant than what has been thought? It is important to take a short look back at some of the things found on the island as to their potential of revealing more than what they appear to be. It could be possible that the searchers on the island may not have recognized the true importance as to any one of those symbols in particular that before their very eyes could hold a hidden cryptic message. It could be that the searchers who have probed the island for decades trying to uncover something in the Money Pit have been too obsessed in their digging operations to realize that there was something else as having more importance. Perhaps it could be the answer as to why Oak Island is there!

Throughout history, symbols have been used by humankind as a simple way of communicating with others in times long before verbal language was shared. Ancient carvings or etchings known as petroglyphs found on the walls of caves left stories that told about the people who had begun early civilizations and their journey through time. Things like the images of the rising or setting sun, or a crescent moon aside a stream with a human figure spearing fish may have told of when it was a good time to fish. Another etching perhaps of crude shelters or huts in a gathering could indicate a village, or a hunter with bow and arrow pursuing a musk ox may have indicated to others a good hunting ground. Those etchings of images translated into symbolic things that had an easy recognition that could be understood by all with simple clarity. As time passed a new style of communication would evolve replacing those simple drawings with different shapes that eventually would become similar to the type of communication used by the Egyptians known as hieroglyphics. Ultimately human societies would begin to adopt their own form of using different symbols and shapes in creating letters. These eventually would be made into alphabets that when seen in an ordered arrangement would create the written words and languages of the earth.

Symbols have become so familiar in our world today that they have come to be understood worldwide by all with the same meaning. Take what is now seen as just a common road sign, the red octagon with the printed word stop or the signage showing a large arrow pointing in a specific direction indicating the route for travel. The raised hand with palm facing outward in one's direction is known as a warning command to the viewer to stop at a specific spot, or sometimes used to indicate a certain action is prohibited from ensuing. The skull and crossbones flown on a flag has long been viewed as the eerie forewarning by those using it as a signal to others that the threat of peril is possible. That skull and crossbones symbol is also displayed on items with labels or posted on signs as a warning of a potential hazard like poison gas or liquids alerting the viewer potential harm is of the utmost concern. Other symbols may indicate love and peace, like the bloom of a rose, the shape of a heart, or the display of a dove in flight. We have come to accept as one of the symbols for Christianity is the shape of two arcs with one being inverted crisscrossing over the other to form the symbol of a fish as is often seen displayed on the rear of many automobiles in today's world. That symbol has a Biblical reference to the story of Jesus when he showed Peter the miracle of faith when fishing. The countless symbols that are all around us everywhere today have an intended purpose in communicating a message to the viewer that some kind of attention is needed or to impart a non-verbal thought, or reaction. So what are the symbols found on Oak Island that need another look?

In one TV episode aired during Season Six of *The Curse of Oak Island,* there was shown a small boulder known as the Evans stone that was discovered with the engraved design of a pole-like tree with a thin trunk. The tree had an odd balance of bare branches on its sides with one side having 17 and the opposite side having 15. Also inscribed on the stone was the date "Aug. 9 1897," and the two names "R W Evans and Chester." The Lagina brothers and others thought the tree design as possibly symbolizing its use during the American Revolution of the 1760s to represent what was called the "Appeal to Heaven" or "An Appeal to God" slogan. The tree that was used in conjunction with the "Appeal to Heaven" slogan was an eastern white pine tree that is usually portrayed as a slender pyramidal shape that had always been displayed with fewer branches than what the Evans stone showed. Pine trees on flags had been used in the New England area of North America beginning as early as the 1650s as it represented the colonists valued use of the pine tree in shipbuilding and types of early housing construction. The slogan "Appeal to Heaven" would become a very important part of American history as a rallying cry when white flags showing a green pine tree were flown atop ships under the command of George Washington at the beginning of the American Revolutionary War for independence. The saying was also used by America's founding fathers during the 1760s as it was also used by patriot soldiers as a plea to God for help in their endeavor to defeat the British. The pine tree symbol, the flag, and the slogan "Appeal to Heaven" were all part of the image presented as a resistance and being defiant to the enemy. It seems curious as to why such a similar etching upon the Evans stone would be found on Oak Island. The stone however had an 1897 date inscribed on it that would seem to indicate the engravings were not reflective of the time period of the American Revolution that occurred earlier in the 1760s.

So what is the relevance of the Evans Stone as to having any clue that we could value as to whom may have done the Money Pit? Since the stone had the engraving of 1897 which supposedly represented the date it was etched, surely indicates an event long after the discovery of the Money Pit in 1795. There is, in my opinion, no distinct tree replication between the Evans stone and any other historically noted tree symbol, past or present, although an interpretive assumption will always be made otherwise by someone. There is a coincidence though of a certain event that did occur on that date which is the death of a Confederate General named Samuel McGowan. The etched names of Chester and Evans may refer as to one of them being the inhabitant of a place known as Chester who was named Evans. Chester is on the mainland of Mahone Bay across from Oak Island. It could be that Evans had worked on the Oak Island project with one of the many treasure search teams when the general had died, and in wanting to honor that date for him, created an etched monument. If that was the reason, why didn't he engrave the general's name? The general was buried in Abbeville,

South Carolina after passing away at his home in August 9, 1897, the date etched on the Evans stone.

There is also one more coincidence about the date of 1897, in that it was the founding date of R. W. Evans & Sons Memorials in Wales, England! A quick search of the Internet can reveal this info in a flash. This is just one more curve ball so to speak about the many things that continue to be a part of the mystery of Oak Island. I am sure there is some curiosity about these two findings, how much impact either have or not could be pursued further. However the dating on the stone seems to negate any real evidence toward solving the actual creation of the Money Pit.

Another known tree symbol during this time period was the "Tree of Life" which had its historical reference that began in the Garden of Eden as described by the Bible in Genesis 3:8,17 and 22. The tree was the one from which Eve had taken fruit that God had described as the Tree of Life and how it represented eternal life. Throughout early history the Tree of Life had been represented in the image of a Baobab tree by the many different religions and cultures of the world who regarded it as a sacred symbol. The tree is different in appearance than other deciduous species in that it appears as growing upside down with the branches spreading out in all directions looking like a root system. The ancient deciduous Baobab tree is believed to have been growing during the prehistoric age with growth of its trunk measuring up to 30 feet in diameter with branches reaching 50 to 60 feet in height and mostly confined to the arid areas of northern Africa. Growing in such an arid region, the Baobab tree was able to produce fruit when others would not, so it became the accepted image of the Tree of Life. One discovery of such tree was radiocarbon dated as being 2500 years old with the belief that they did live even longer. The image of the tree's branches and roots were seen as being representative of the different significances to the existence of humankind as to how life is organized with everything in the universe being connected. A historical find of the Tree of Life that had been etched in stone was found at an excavation site in Turkey dating it as far back as 7000 BC The tree species still exists today in some areas of northeastern Africa but is threatened by recent climate changes. There are numerous artistic representations of the Tree of Life that can be found in paintings, jewelry, and geometric designs that are used by the various religions, each with their own unique interpretation assigned to its' various parts.

The actual biological image of the Baobab Tree of Life has been represented in modern times by a complex geometric construction consisting of 10 different points representing 10 different meanings. The geometric design is similar to a tree with the bottom point or node being the trunk with branches rising upwards to the crown. To give details of how each part of the Tree of Life has been interpreted by all can get rather involved explaining each of their differences and would require much space here. A brief inclusion about the Hebrew interpretation involving what are known as the ten sephiroth can though be included. The Kabbalah, an ancient Jewish tradition has interpreted their beliefs in the Bible to identify each of the ten sephiroth (sefirot) as the powers by which God has manifested. The geometric form of the Tree of Life comes from the belief that geometry reflects the perfect

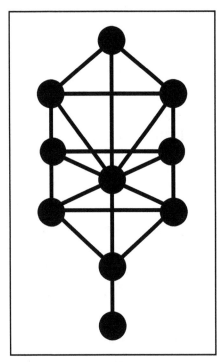

Tree of Life Symbol

order of nature in the universe with sacred meanings inferred to each part. There are other geometric shapes similar to the Tree of Life as in the one that uses the planets and the sun at its center with each depicted as a sphere on the tree.

Theorist Peter Amundsen appeared in an early *Oak Island* TV episode, and described the Tree of Life geometry as being hidden in the design of Nolan's Cross.

The Liberty Tree was another familiar symbol used during the American Revolution as it stood to represent the idea of revolution the American colonists felt leading up to the War for Independence that began in 1775. The Liberty Tree was a species of deciduous tree belonging to the elm family that branched out in many directions standing as tall as 60 to 80 feet in height with a wide trunk and a spreading top or crown making it shaped like a light bulb. How the elm tree became the symbol of liberty arose from the protest events leading up to America's War for Independence that were held around a particular 120-year-old tree that was located in Boston. The protests were led by a group of nine colonists who were known as the Sons of Liberty and were opposed to the British Tax Act enacted in March 1765. The tax had been imposed upon any documents the colonists created, such as deeds, mortgages, newspapers, licenses, and the like. The Tax Act along with a tax on sugar had irritated the colonists to the extent that it would lead them to do many protests that were usually organized by John Adams. Those protests were held around the old elm tree. As a warning to the colonists, the old elm tree was cut down by those who were loyal to the British crown as a message to those would be non-conformists. The fate of the tree inspired the cry for revolution to obtain liberty!

There are other tree symbols that could be discussed, but none that would match what was etched on the Evans Stone. In short there is no certainty as to what the etching was referring to, any opinion is as good as the next.

An area of special focus by the Lagina team has been the so-called swamp. It is located between the two land masses of Oak Island that form a large triangular shape that is thought to possibly be manmade as a hiding place for a sunken ship. This site has been the subject of much past and current speculation as to its triangular shape as having some special purpose of importance. It seems by all who have been involved with the swamp area, especially the late Fred Nolan, that it must have been man made with some important design in mind. There is ample suspicion to believe it was manmade centuries ago by those yet unknown. The swamp has created but one more mystery yearning to be solved. All of the speculation about the swamp having been manmade though is up for debate as it could have been formed by natural occurrences done over many years by the ebb and flow of tidal forces that cause erosion and sediment infill. If it was naturally made, it was a very good geometry feat done by nature as its shape is almost a perfect equilateral triangle!

Lately there have been some underground abnormalities discovered about the bottom of the swamp with the use of modern technology. Special types of sonar equipment have been used to display graphic recordings of underground geological irregularities that have indicated an elongated large shape buried deep in the swamp. Could it be a sailing ship of yesteryear as some have speculated that was possibly scuttled on purpose to hide it from discovery? If it is a ship, why was it put there? Could it be laden with rich treasure intended to be secreted away from all to never find? There have been some discoveries of artifacts found in an around the swamp area that are normally found as being parts of a ship's structure such as a few very old wooden decking planks and handmade iron nails and spikes, typically used in ship building. Even a recent find of a once buried roadway along the side of the swamp made of stones and small rocks leading from the shoreline inward toward the Money Pit has created an interestingly new mystery. These things do cause reason to believe that the swamp is more than just a swamp site with definitely more to be told, but what? The swamp, in its triangular shape, certainly has a mystery unto its own that we need to understand. Only time accompanied by much more searching in the depths of the murky swamp will ever give us any glimmer of hope in knowing what is confined within, or why it was done.

The equilateral triangular grouping of stones found near the south shore area of the island that had sides about nine or 10 feet in length probably indicated something of importance, but what? The stone shape was found in 1897 by Captain Welling who was one of the earlier searchers, and it was later examined in the early part of the 1960s more closely by Fred Nolan, who photographed the shape prior to its destruction in the mid-1960s. The stone collection was determined to be similar to the shape of a sextant. The sextant had been used as an early navigational tool for measuring the celestial angle of the stars in determining distances and direction for travel. Mostly used by seafaring navigators and ship captains, the sextant was first used in the late 1600s gaining more widespread use in the 1700s. The stone configuration had a line of stones starting at its east to west base that were pointed northward through the center of the figure to the apex of the triangle. The shape was undoubtedly pointing toward something which was later found to be a large boulder near the Money Pit! Any of the searchers involved after the discovery in 1795 could have left that stone configuration as a reference for later users to easily be directed from the shoreline to the large stone found several hundred feet away near the Money Pit. Since the shape was that of a sextant which was commonly used to determine directions and locations by ship captains and their navigators, it seems very likely that some seafarer was the designer of the stone configuration. But who? One more question to answer.

Most of the time when a triangle is seen, the implication of how it is placed is for the purpose to point in a particular direction to travel or to look. The triangle shape also has its own symbolic meaning in the world of spirituality as being the symbol with connections to the omnipotent existence. The triangle has been referred to as the three essential elements of the human being: the mind, the body, and the spirit. It is also devoutly used in Christianity as to mean the Father, the Son and the Holy Spirit with each entity being equal just as the sides of the equilateral triangle parts are equal combining to make one complete shape as does each separate religious entity combine to embody one God. The importance of the triangular shape still needs some more inclusion that will appear again in later paragraphs, as its geometry does become a prelude to the final clue.

Nothing so far has led to any conclusive evidence that can be considered one hundred percent as any undeniable evidence of who did what on Oak Island prior to 1795! So we need to keep examining what we know so far that is physically present on the island that can lead to a possible answer. Something is still eluding those looking for that one connection to the truth about what is being sought. If there was at least one good solid lead, then all could enthusiastically shout, "This is it!" What a tremendous breakthrough they would achieve. Sure, there have been some clues that have become very helpful, but even the ones that have been made note of so far have not been able to put a joyful "Ah Ha!" into Rick Lagina's vocabulary!

Are there any other shapes or symbols on the island that can lead us to a potential clue about what is Oak Island's secret? The uncovering of some structures that seem to have been constructed for use either as piers for the unloading or docking of ships have only left the searchers with more questions as to why they were built, again leaving the questions of by whom and when. Just another mystery upon so many other mysteries that have already become too entangled to unravel. So is there any clue at all that can help as to who left any evidence on the island that has created one of the world's most intriguing mysteries! There is only one more physical artifact to examine!

It is time to look at the cross that Fred Nolan discovered many years ago during his surveying activities that led to its discovery. It is the symbol recognized as being the universal sign of Christianity ever since the days following the crucifixion of Jesus. Why was it found on Oak Island, constructed on such a large scale that used massive boulders in its design that were very heavy with each weighing several tons? The boulders were also placed many feet apart and were laid with extreme accuracy in a very precise geometric configuration? There had to be an extraordinary reason for such an undertaking to occur! It must be acknowledged that an event like this could only be accomplished through the Herculean efforts of a very large and powerful labor force! Who could that be? The ingenuity that was required to perform such a feat had to be orchestrated under the direction of a very capable engineer with exceptional creativity!

Not too long ago after being hooked on the TV series of *Oak Island*, I began to look into what might be the intent of creating such a huge boulder configuration known as Nolan's Cross. I had taught math in a public school after retiring from real estate sales, and geometry had become my favorite subject to teach. As a realtor, my hours working outside the home during evenings and weekends were not what you would say compatible with a teacher's work week. Teaching math in a public school setting renewed my motivation about mathematics and I needed to hone some old skills once again. I always had loved the field of geometry and even studied architectural design at one time in the hope of becoming a drafting engineer, but life throws a curveball sometimes as things change. I did use a lot of math during my time as a realtor that involved geometry to figure out building floor plans and land measurements. Anyhow, my experience with math and especially geometry had peaked my interest in wanting to explore what I determined to be the special construction of Nolan's Cross. There was something puzzling to me about why the scale of the cross was of such a large configuration with such special dimensions.

The cross has its own unique identity that gives it a special place in history, that is not denied, but why was this creation resembling that of a Latin Cross found on Oak Island? Most people would say that this style of a cross means only one thing, it is the symbol of belief that Christians have in the teachings as written in the Holy Bible. So it would suggest to anyone aware of the Oak Island cross to believe it must be there for some reference to Christianity. Could this symbol be the clue that could unravel the Oak Island mysteries? Is this the clue with a message that hasn't been understood? What kind of message other than the obvious one that it's a cross known worldwide by Christians? We must investigate beyond the obvious because there could be something that has possibly not been realized!

There are other religious symbols that may help us to better understand what we are looking for on Oak Island. It is how a few of these other symbols each contribute a piece to the significance of why the unique configuration of the cross is on Oak Island.

Let's go back in history to explore how other religious symbols were used even before the Latin style cross was used to signify the Christian belief. The importance of why differing symbols were used by differing religions was to give an immediate recognition of what religion that particular symbol represented. But some symbols had similarities to others with like meanings. For the purpose of this book, the symbols most relevant in history are the Star of David sometime referred to as the Seal of Solomon, the Latin cross, and the five-pointed pentagram or star.

King David was the second king of Israel and is most known by his defeat of the fierce giant Philistine warrior named Goliath. David lived from 1035 BC to 970 BC and was a faithful believer in God with traceable ancestry connecting him to Jesus of Nazareth as written in the New Testament of the Bible. A part of the Judah tribe of Israelites, David was the youngest son of Jesse, a shepherd. David also becomes a sheep herder in his youth learning the skills needed to protect the herds of sheep he oversaw. His protective skills would become honed like those of a skillful warrior. One incident that describes his fighting abilities is found in I Samuel chapter 17 verses 34-37 where David once kills a bear and a lion to protect his sheep, then gives praise to his faith in God that enabled him to have the strength and skill to do so. The Israelites were enemies of the Philistines, living in the lands next to each other they would have constant mortal conflicts. On a day when the two armies prepared to do battle once more, David is present along with some of his brothers to witness the giant Goliath emerge in front of his troops to challenge the army of the Israelites to combat. Offering a compromise, Goliath dares the Israelites to a one-on-one battle with the victor to claim all of the lands. Seeing the enormous size of Goliath at 6 cubits in height (about10 feet tall) all clad in armor with a spear the size of a small log, the older skilled Israeli warriors shuddered with fear, knowing they would be easily defeated. King Saul, the first king of Israel and leader of the army, is approached by David who claims he has no fear of such an adversary as God is on his side. The King accepts the bravery of the younger son of Jesse offering to equip him with armor and shield for protection, but David refuses declaring that God is his only necessary protection. Saul makes a vow of riches with a place in the house of the king for David if he becomes victorious. David goes forward to confront Goliath as the warring faction of the Philistines approach the lands of the Israelites to do battle. Goliath scoffs at David as a puny adversary, but David's unyielding belief in God and his accuracy with a slingshot sent a stone to the forehead of Goliath bringing him down to the ground in death. Why David confronts such a formidable enemy while other Israeli fighters were unwilling to do so is because of his faith. The strength that David had shown in his exemplary belief in God had proven to the Israeli people that good can outweigh evil even when the threat may seem unbeatable. This Bible incident is just another example of how conflicts that involved the Jewish people enabled them to defeat enemies through their absolute belief in God. Some have said that God was the shield that protected David in the many battles that he ultimately would fight in protecting the Jewish people in the name of God. The Shield of David has also been referred to as the Star of David.

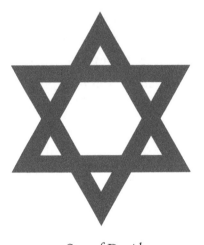

Star of David
Public Domain Image

King Saul would eventually die after suffering an illness, leaving the next throne occupant to be elected by the will of the people who name their young hero David as the new king. The biblical story of David is very interesting and should be read; what is important for this book is what King David did next.

God sends a message to King David that he is to make arrangements for a permanent structure to house the Ark of the Covenant. The Ark of the Covenant had always been sheltered in a temporary tabernacle which is similar to an ornate tent structure that was constantly having its location changed since its inception. Wars among the 12 tribes of the Israelites and other enemies had not afforded a peaceful nor permanent safe site upon which to worship the Ark, so in order to protect it, ease of mobility was a necessity. David does gather all of the necessary materials to build the temple to house the Ark, calling on his friend, the king of Tyre, named Hiram to help with the needed labor. This has all been described in an earlier chapter in the construction of Solomon's Temple. What I want to do here is to emphasize how symbols like the Star of David or the Shield of David begin to influence the importance into what is to be found on Oak Island.

As best that can be determined, there is no reference to the hexagram symbol as the Star of David in the Bible! Let that statement be repeated again, there is no reference in the Bible to the Star of David. There is also no mention of the Star of David symbol also known as the Magen David (shield) as being a geometric shape in the Torah as acknowledged by many religious academics and Rabbi's. There is though wording that mentions the divine power or protection of David as that being his shield and is found in the siddur, the daily prayer book used by the Jewish people. However the geometric hexagram symbol has only been recognized by the current Jewish state since 1948 after it was used in the second World War of 1941-1945 by Germany to identify Jewish citizens. The symbol was chosen by Israel to become their national flag in 1948 after declaring Israel a duly recognized state. The flag displays the six-sided hexagram which is created by using two interlocked equilateral triangles in the color of deep blue centered on a field of white with a wide blue horizontal bar at the top and bottom of the flag measuring five feet by seven feet. Since the symbol of the Latin cross had already been used by Christians for many centuries to signify their faith, the symbol of the hexagram representing the Star of David, the Jewish ruler, was very appropriate in distinguishing the uniqueness of the Jewish faith. Each of the 12 major religions in the world do have differences in their beliefs, as does each one have its own exclusive symbol, however, they all subscribe to a belief using their own interpretation that there is a supreme power!

Most research indicates the Jewish people began using the hexagram symbol as a form of a design that was repeated early in their culture. It was found on the facades of synagogues and homes as well as on ornaments of other Jewish artifacts, and it was used as seals and seen on writings. The shape of the six-sided star has been discovered in Galilee on the ruins of an excavated synagogue that was dated to the 3rd or 4th century seemingly used as a symbolic decoration or design. An earlier example of its use was discovered on a seal recovered from the ruins of the ancient city of Sidon that was destroyed in 675 BC. Sidon was located in Lebanon on the coastline of the Mediterranean Sea. The shape itself of two equilateral triangles that are superimposed over each other to create the six-pointed star has been found in history beginning as early as the Bronze Age dating between 3000 BC-1200 BC. During the Bronze age the symbol was thought to have been used as a tool to create magical powers over demons. The geometric design was believed to have been done to indicate one of the triangles that points upwards as good while the other pointing down as bad just as heaven is to hell with each pointing in opposite directions. It has also been interpreted as the upward pointing triangle meaning the soul or existence of man with the downward pointing triangle referring to that of the woman, the two opposites of humankind present on earth. Each triangle pointing in a different direction is also believed to represent the balancing of any opposites thus giving harmony to whatever the applicable concept, like the yin and the yang as in Chinese philosophy.

Another interpretation of the star symbol is that each point represents the six different directions of east, west, north, south, up, and down and that all are under the eternal control of God the omnipotent one for infinity and beyond!

Solomon was the son of David who is attributed to the building of the temple in Jerusalem that housed the Ark of the Covenant. He is also known as the wise king who made very logical choices that helped Jerusalem flourish while also uniting all of the Jewish tribes of Israel together. What has been described in an earlier chapter about the events of the temple should suffice as to its further history not being necessary here. What is needed here about Solomon is the legend of his use of a magical ring. The ring was considered to be the Seal of Solomon and it could control evil or good along with some other powers. The ring was described as a type of signet used as a seal or stamp that was worn on the finger. The ring was supposedly made of iron with some brass that displayed the geometric hexagram design as constructed with the use of two equilateral triangles being interlocked facing in different directions confined within a circle. The ring, as is believed by some Arabic writers of medieval times, was believed to have received magical powers that were sent down from heaven by God.

The geometrical six-sided hexagram formed by the two interlocked equilateral triangles to form a star, was not only found in the lands of Israel, but was also used in earlier times by those of the Hindu faith. One of these geometric displays used by the Hindu is the encircled hexagram symbol used to represent the human heart which in Hindu is called the Anahata.

Hinduism began sometime around 5500 BC, long before Christianity in a region of India. The belief is not restricted to worshipping only one god as in other religions, but acknowledges many different gods with a teacher, leader, or knowledgeable one at the head of all the teachings called a Brahmin. Initially Hinduism began as a tradition of values that were orally described in stories up until about 1500 to 500 BC when they began to be changed into written form and were copied throughout the lands. In the Sanskrit language of the Hindus the word "Anahata," or heart, has the meaning of "unstuck" and being "clean" or "pure" with unhampered devotion. Anahata is symbolically represented by the likeness of a blue colored lotus flower that has a hexagram (like the Star of David) inside a circle at its center. This particular charka (circle/wheel) representing Anahata also symbolizes love and calmness and is one of the seven chakras used to represent the different areas of the human body. There are four more chakras that also use the inverted triangles within the circled wheel that identify other parts of the human body. In Hindu the two inverted triangles are referred to as the Shatkona, one triangle pointed up (Shiva) as representing male, and the downward one (Shakti) as being female. Here again the opposite inverted directions of the triangles have the same meanings as are used in other cultures. This theme of the up and down pointing of the triangular symbols representing each as the opposite of the other, has been continually repeated by most other religions and cultures throughout history.

The use of triangles to form a six-sided hexagram shape by the Hindu could be with the intention that it also helped to represent their beliefs such as in the six faces of Shiva & Shakti joining as two for the purpose of creation. One Yantra known as the Sri, or Shri Charka has a total of nine interlocking triangles that represent the powers of the universe, a very powerful and respectful acknowledgement to a higher entity.

Other geometric symbols are also used by the Hindu in creating their 56 different worship Yantras which are colorful, mystical and very artfully designed diagrams, each with its own significance. What is evident here again is the importance of the triangle as a symbolic religious shape and how its use by the many cultures throughout history has echoed the same in its similar representation of a belief.

Islamic use of a religious symbol was adopted during the Ottoman Empire, which existed for 600 years between the 14th and 20th centuries beginning with the use of a five-pointed star and crescent moon as a flag to represent their many different nations. The star can either be thought of as the "morning star," and the crest that is shown as the "waning moon" phase. In the 20th century the crescent moon flag has become an exclusive symbol for the Islamic

Symbol of Islam
Cresent Moon & Star
Public Domain Image

Arabs and is currently used as the flag of Turkey. The flag however is not openly accepted by all of the Muslim faith.

The importance of this segment is to acknowledge the use of geometry and geometric designs as they are used by the many different religions to present certain symbols as having a sacred identity.

Looking at the various religious symbols presented so far, which are often displayed as flags, jewelry, or in print, are certain geometric shapes that are commonly shared, such as triangles, stripes, parallelograms, stars, pentagrams, crosses, circles, and curves. It is not unusual to see several of these shapes combined to create a unique motif as some that were discussed here, and when they do, some beautiful and creative geometric artistry awakens the eye of the viewer. It is evident that these various shapes were used with a resolute religious intent to signify certain shapes, or their combinations as representing a very specific religion, even though some may have shared similarities. Each different religion however has used their own unique, but very identifiable presentation in a special creative design to set their symbol apart form others. How the geometry of each shape was used to design that symbol to be unique unto itself was done with the utmost caution as to not replicate its use, or not to use the same shape in an offensive way toward other religions which could provoke a religious war! With that in mind, there is one shape used by many that has no specific religious ownership. That shape is the five-pointed star or pentagram. There are in today's world about 90 different countries with a flag that display at least one five pointed star symbol used as a part of their national identity, but not necessarily used as a stand alone religious image as do the images of the Latin cross. However, there are examples of flags using stars to represent a specific religion like that of the Baha'i in Iran which uses a nine-sided star as a symbol. Baha'i has just recently come into practice in the Middle East during the 19th century. Followers of Baha'i separated from other religious groups in the Middle East which is most often

Knights Templar Battle Flag
Image courtesy of Hammerson Peters of Mysteries of Canada

the reason for creating a uniquely identifying symbol unto itself. Islam, also a religion of the Middle East, uses the five-pointed star, but it is used in conjunction with the crescent moon as seen here in an earlier description. Another modern day religious use of the five-pointed pentagram star is found at the Church of Jesus Christ of Latter-day Saints that is displayed on their temple in Nauvoo, Illinois. The star symbol can also be seen at the other temples of the Latter Day Saints in Salt Lake City and Logan, Utah. Their flag uses a field of blue with white horizontal stripes showing one star encircled by 12 smaller stars representing the original 12 tribes of Israel in the upper left quadrant of the flag.

With the use of the star often seen in the Middle East regions, it was not a surprise to see the star appear on a Knights Templar battle flag they used during the Crusades in the Holy Lands. One of the Templar flag designs is interesting in that it also displays the crescent moon of the Islamic flag. The most dominate part of the flag is of course the red splayed Latin cross on a field of white. Other Templar battle flag designs used were ones displayed with evenly divided rectangular horizontal fields with one in black over one in white with a

Knights Templar Flag

large "la croix pattée," the red cross centered vertically over the two parallel fields; that flag was called the Beauceant. Not only were the Templar flags displayed in the Holy Lands, but as will be noted later in this book similarities were also found in other far away lands, and certainly displayed on Templar sailing ships as well! Looking back at the ships Christopher Columbus commanded, the *Nina*, *Pinta*, and *Santa Maria*, a very similar "la croix pattée" can be seen!

The use of the image of a star symbol or pentagram, has been dated back in ancient times as early as 8000 years ago, and has been used by many different human cultures to signify varying things of importance. It's geometric design has been incorporated in representing certain religious, or sacred beliefs, or presented in a concept of something that was viewed with the very highest esteem, or reverence. It has been found on artifacts like that of the domestic pottery found in the ancient Holy Lands that were identified as being used by those of the Hebrew or Jewish faith. The star symbol was also known to have been used by the Phoenicians who were known friends of King David, the first ruler of Jerusalem.

So, is there some geometric image, or symbol that is of historical significance to those found in the many different cultures, religions, or belief systems around the world that could fit into the mystery of Oak Island? Yes! But what is it?

Important Things to Consider

Throughout the history of humankind, symbols have been used to tell stories beginning with the paintings and drawings found on the walls of caves around the earth. Left by our earliest ancestors dating back hundreds of thousands of years, examples like the sun and the moon were used to symbolically tell of the passage of time. The drawings of hands on a wall told of family groups, skulls and hearts told of death and love. Animals, fish, weapons, geometric shapes of circles, rectangles, patterns of dots and lines, lightning bolts, and other shapes were all describing life through a symbolic picture story. The Egyptians developed hieroglyphics, the beginning of a structured language that used drawings of living things like animals, plants, birds, humans, and other kinds of earthly things and included many strange shapes. Their language has been catalogued to include over a thousand different characters that were used to communicate among themselves and with others while leaving a story in an ordered form about their history. The Phoenician civilization in the 8th century devised an alphabet using twenty two symbols in various fixed orders to create words that were quickly accepted throughout the Mediterranean as an easy way to communicate in written form. It would not be long before the Greek empire would utilize the Phoenician technique and began to use their own version of ordered symbolic shapes to create words. They then advanced that style to write words in a sequence and began to structure sentences by writing words from left to right, the method most used today. All this gave birth to what we recognize today as the English alphabet, using 26 symbolic letters to form various words and sentences.

What humankind has done in creating alphabets have found a way to interpret these symbols into familiar letters organized in such a way as to create written communications that result in evidence of a message. Some alphabets use different looking symbols to represent the same meaning as in other alphabets. Chinese symbols or characters are different then Arabic, and Arabic is different than those in the Russian language, and so forth. But all communicate using the same technique of shapes that make symbols called letters.

Symbols in one form or another are used by humans to communicate a message. Some symbols can be in letter form to create words and sentences, or others like a red octagonal shaped sign at a road intersection, should be interpreted by the driver of a car as a need to stop. The image of two intersecting lines forming juxtaposed arcs with extended tail ends crossing is called an ichthys, the Greek symbol for fish. It is also the early symbol used by Christians as a reference to Jesus Christ, and is seen today displayed by many Christians on their personal possessions. The Latin cross is the symbol of Christianity, recognizing the structure that Jesus Christ was crucified upon and is considered a holy display of faith when observed. What will be described in this section of the book is that there have been messages left throughout history in many symbolic forms. Some of these messages have been obvious and meant to be understood, such as an alphabet that forms different words. Some messages were designed to be seen, but are only meant to be understood by those who know their purpose. Art work, in its various form, is a message presented by the creator that most often can be seen as being very interpretive to the viewer in communicating an understandable message. A painting of a ship at sea in a raging storm presents an image to the viewer as a probable perilous situation. Artists have created certain works that can be seen as almost lifelike, while others paint or model images in the abstract, leaving the viewer to create in their own mind what is to be seen. Paul Klee, a very famous German abstract artist, paints shapes in all various forms and sizes using various colors displayed in erratic configurations. His art is left to the viewers imagination, but he does convey a message that can be deciphered. There seems to be a message on Oak Island that has been left by those with a secret, for no one knows why an array of boulders forming a cross that spans many feet is there. The cross was recently discovered after being unnoticed for probably centuries. Surely it was not meant to be obvious by the way the designer spaced the identifying boulders so far apart. Even if someone does see it now, there must be some kind of special meaning for its presence. Is there a cryptic message that must be analyzed first to understand the intended purpose it presents? Like a piece of abstract art, or a new alphabet, how do we interpret it?

To understand what the objective of any given message is intended to accomplish it is necessary to know what is the guiding purpose toward its presentation. For example, if you see a billboard along the side of a roadway that shows an outreached human hand, it can be assumed the message is one of need, or a hand of help. But, if at the bottom of the billboard you see the wording: "To donate, go to www …", you do understand it is a request for need. The end goal of any message, whether it uses symbols or words, is to deliver a form of thought originating from one source and given to another that can be acknowledged and hopefully understood. The highway crew put up the stop sign symbol because there is a need to stop at that location, and the vehicle driver understood it's a message, and stops. When someone creates a message, whether with a symbolic structure, or letter characters in forming words and sentences, it must have a purpose for a desired end.

The discovery of the layout of some very large boulders on Oak Island that when mapped, form the image of a large Latin cross, a very symbolic shape representing Christian ideology certainly must convey some form of a message. The engineering and precision required for such an undertaking would be a monumental task not done without the intent of a very special purpose! But who put it there? How should the cross be interpreted? Is there a connection to some part of history that the message of the cross is wanting to tell? The answer will be told herein!

To begin to answer why the cross is on Oak Island, one must focus toward the thought of what might be the reason why it was put in that particular location. The cross is a form of communication about a message that has a purpose, and was in all probability, left there as a sign to identify that spot as an important part of the Oak Island story! The story must be of significant importance for those who left the sign there, as they went to great efforts using forethought in being so precise in its geometric construction.

Let's begin with why a cross? Why not another symbol? Maybe a huge ship's anchor that would suggest the landing of a large naval flotilla. The island is located where many sea travelers venture from many nations, and could possibly indicate a certain place to call a safe port for those who sail the surrounding waters. How about a giant X instead of a cross? The X marks the spot could be easy to construct, and has been used for centuries to indicate the location of something to be found, like buried treasure on an island with the X drawn on a map. A skull could imply the thought of the island being deadly, as it truly has been for several souls who have searched for something unfound. Even a pile of large boulders or stones all heaped together in one spot, similar to a small pyramid have in olden days been used to identify a special location. Any of these stone configurations could have been used on Oak Island to convey an intended message, but instead, what has been found is a cross! Why?

The answer to why a cross is on Oak Island, undoubtedly suggests that the symbolic Christian shape is a special message which has been purposefully left to indicate an evident Christian purpose. So if the message does have a Christian purpose, then who is the originator of the message, and why is it there? We must search the beginnings of the usage of the cross and any other related symbols to try to understand first who would use a cross, and may have placed it on the island. In an earlier chapter there was the examination of past theories about pirates and privateers having been the originators of the cross on Oak Island. Although it was noted that some pirates and privateers were Christians, they likely did not have the opportunities to manage such a labor intensive task without being physically discovered by others, or eventually being exposed by some untrustworthy participant of the project. There is definitely no record of any such event ever being recorded by anyone, much less being acknowledged by pirates. Pirates were known to turn on each other for their own selfish gains, and any secret that was meant to be kept and never to be told would have only lasted as far as the next port of call. Let's rule out the pirate connection to the placement of a Christian cross configuration on Oak Island. Pirates may have visited the island to stash some treasure, but in my opinion, had no purpose in fashioning a site for such a huge cross.

There are other suspects that could be considered, such as the early English, or French explorers, maybe even an occasional seventeenth or eighteenth century military operation seeking to establish a new territory for their home country. There is the possibility that one or more of the early naval military operations could have visited the island to stash a fleets' treasury for temporary safe keeping, but did not have a reason nor the opportune time to construct this elaborate cross. The cross could have been done by the settlers on the mainland of the continent as a way to claim the island as a Holy location for worship. Yet, here again there would be some form of recordation by a local historian, or at least some local lore of them having participated in such an activity on the island.

It is very doubtful that the indigenous Mi'kmaq tribes had any knowledge of early Christianity prior to any European influence which would prompt them to use the form of a Latin styled cross. The Mi'kmaq that were found throughout the northeastern region of the eventual Canadian territories, including Nova Scotia and Newfoundland, were believers in the sprit of Kji-Niskam. It was their belief that the earth and living things were all of one spiritual presence created by one being and that life and death was an eternal circle. Some Mi'kmaq would eventually accept Catholicism in the 17th century after it was introduced by the Europeans, but their belief in Kji-Niskam is still worshipped today. Although the Mi'kmaq had access to the island and could have possibly engineered the movement of huge boulders to create such a complex array of the cross, they seemingly had no purpose to do anything relating to a Christian Cross during any time prior to 1795. The Mi'kmaq would however adopt the use of a cross and crescent moon opposite a star on their flag that was adopted later in the 1980s that was similar to the Knights Templar flag! I do need to refer here to something stated by some that the Englishman Sir Henry Sinclair, a possible Knights Templar did come to Nova Scotia in 1398! The Flag of the Mi'kmaq is actually the mirrored version of one of the battle flags that the Templar have flown! This is just another indication of the Knights coming to North America! To review the flag see the previous chapter, "Symbolic Clues Leading to the Answer." A good question here is why did they adopt a very similar flag to the one used by the Knights Templar? Another mystery about the Knights Templar in the area of Nova Scotia, or should we say clue!

Is there any evidence to suggest who or why the cross exists on Oak Island? Yes! There have been some items found during the many decades of searching that can lead to a strong clue as to the identity of its originator! Almost every searcher involved has discovered something that could steer the discovery toward knowing who and why. At present the most successful searchers have been the Lagina brothers, their partners, and other supporters. Their findings resulting from their searches and their investigations into the historical connections have brought them pretty close to getting the answer they seek. But close is only good in playing horseshoes or throwing hand grenades as the old cliché goes.

There is the need to look at what has been literally uncovered so far that could be some type of a message that has just not been interpreted. One of the first very mysterious items found was that of the 90-foot stone, discovered in the summer of 1803 or 04. The stone was described as a slab measuring two feet in length, about fifteen inches wide, and almost a foot thick, having carved geometric symbols and strange letter shapes that could not be deciphered. Unfortunately, over time, the stone has since disappeared without anyone ever having any real idea of what ever happened to it! Some did try to interpret what the written code was before the stone disappeared, concluding that it read, "Forty feet below two million pounds are buried." This item does not help in giving any clue as to the purpose of the cross. The only relevance that the stone has, is it was not similar to any native stone types found in the North American region! It has since been determined by using historical knowledge about what the stone resembled is suspected to have come from the northwest regions of Europe! That fact alone presents a clue or message that the originator of the stone, and probably the Money Pit, did come from northern Europe!

One more theory about the 90-foot stone engraving is that it is believed to have been done as some type of directions in cryptic mathematics. That could be a possibility, yet no one has ever decoded the symbols to be interpreted into anything other than the previous results. There is one clue that can be taken from the interpretation that may be important, and that is the use of the implied word of pounds. Pound has been used in weight measurement since the Roman era as a way to measure the exchange of goods as barter in the Mediterranean areas around the 8th century. One pound of weight has in the past been based as the mass equivalent to 7000 grains of wheat and is divided into 16 equal parts, or ounces. The measurement used for coinage was slightly different using 5400 grains called the tower pound as in the Tower of London, that equaled only 12 ounces. The terminology of the word pound was widely adopted by the Europeans in the 13th century after the Avoirdupois System of weight measurement was introduced in England.

The English sovereign was a gold coin introduced in 1489 by Henry VII, and was sometimes called "the pound" based on the value of it being worth one pound of sterling silver. I can imagine some old pirate saying "Aye, a pound of gold for a barrel of rum." An amount of gold weighing a pound would be quite a price for a barrel of rum, whereas a gold coin worth a pound of sterling silver would be different. Sterling silver is a harder more durable silver in that is combined with 7.5% copper, a slightly harder metal. The use of pure silver for coinage was found to suffer quicker wear, thus becoming easily defaced early in its use. Silver coinage during this era in time was used mostly to mint smaller denominations such as the English pence similar to a U.S. penny. The first gold sovereign coins were not really used for circulation as legal tender, but more as a gift from King Henry VII to people of importance. That coin was made of 95.85% gold weighing about half a troy ounce and larger in size than a current U.S. silver dollar. After the year of 1604 gold English sovereigns were not minted again until 1817 with a smaller size coin weighing only one fourth of an ounce.

Since the re-introduction of the English sovereign coin (pound) it has been devalued many times through the centuries due to the rise and fall of currency values and the establishment of the gold standard. Depending upon which interpretation of the pound the 90-foot stone represents would certainly affect the value of any treasure. Is the inscription on the stone talking about the coin called by some the pound, a gold sovereign of old, or is it describing the weight of some treasure like gold metal? If there is two million pounds of gold metal buried on Oak Island, based on the value of gold in the year 2022 of almost $1800 per ounce at 16 ounces per pound, the treasure would be worth about $57,000,000,000! That would be a pretty substantial treasure! Should the treasure not be gold, but instead would be two million pounds of silver, then at today's price in 2022 it would be worth about $650,000,000! Still a pretty sizable treasurer to find!

Using the other interpretation of it being two million pounds of old English gold sovereign coins (pounds), the calculation with numismatic value being added would be an even greater amount becoming an astronomical figure! According to estimates by numismatic experts, the value of the earliest gold sovereigns could run into the 1 million dollar range plus per coin! Every gold coin collector in the world would cherish the ownership of any value gold sovereign as they are the most collected gold coin in the world!

If there are gold sovereign coins on Oak Island, then they all would have dates ranging between 1489 and 1604 the last year that the original size was minted. Any more sovereigns were not minted again until 1817, 22 years after 1795 when the trio of Smith, Vaughn, and McGinnis discovered the pit!

Depending on the decoding of the 90-foot stone as being correct in either of the actual meanings of the word pound, the treasure to be found on Oak Island could be the most valuable find in all of history! If that is the treasure for the Lagina team to find, they are going to be incredibly rich!

Let's assess the "two million pounds" as being in the form of gold metal bars. The type of vault needed to contain two million pounds which is 1,000 tons of gold, would have to be constructed with some pretty substantial building materials especially if it was to be buried underground in the so called Money Pit. Considering gold bars are usually formed in different sizes, it would only be an estimate about the size of the possible treasure buried on the Island. In today's market the size normally used for gold bars that might be similar to those used back in the 1300s or later would be about 400 ounces each which is 25 pounds. Each bar would measure approximately 10 inches in length by three inches in width by 1½ inches thick. The 400 ounce gold bar is today most often referred to as a Good Delivery Gold Bar. The term "Good Delivery" is used to describe gold bars that meet the rules and regulations of quality control standards set by the London Bullion Market that are accepted worldwide for trade on the open commodities market. In calculating the space that two million pounds of gold bars measuring the size of the Good Delivery Bar would occupy, will first require some calculations of how many bars at 25lbs. each to equal the weight of two million pounds, or 1000 tons. I am not going to go through all of the math combinations that were used to come up with something that seems probable to what this pile of gold bars would look like. Lets just start with the simple fact that the number of gold bars in 2 million pounds divided by 25 pounds for each bar, would compute to the incredible amount of 80,000 bars! That fact alone is awesome! Now take the measurements of a usual single bar that is 10 inches by three inches by one and a half inches and begin to start building a stack of bars that would be manageable. I used the length time width times height of 25 x 40 x 80 to get a stack that would be 80,000 bars. That calculation is 25 x 10 inches, 40 x 3 inches, and 80 times one and a half inches, which computes to about 21 feet in length by 10 feet in width and 10 feet in height. That is one big stack of gold bars!

Should the interpretation of pounds be applied to the treasure buried as being silver instead of gold, then 2 million pounds of silver in today's market would be worth about $730,000,000! Still a pretty substantial treasure to find!

The size of the area needed to hide all of that gold would look like a small warehouse or a big storage shed. Could the believed Vault at the bottom of the Money Pit be that big?

The number of ships it would take to carry all of that gold would almost be like a small fleet. The pirate known as William Kidd a Scottish privateer who was rumored to have sailed the waters around Nova Scotia captained a ship named the *Adventure Galley*. That ship was only 103 feet long and about 25 feet in width. A ship the size of the *Adventure Galley* was usually rated to carry about 284 tons or 568,000 pounds, surely that would not be adequate enough for carrying the entire estimated hoard of gold. The limited hold dimensions below deck of a ship this size would probably only accommodate a small portion of the estimated 80,000 gold bars without making several trips. Other areas below deck were usually limited in space like those areas where the crew members found their quarters. These areas were sometimes so limited that crews could only walk in a hunched over position because of the limited headroom between deck levels. Other areas below deck included the cooking galley and room for the ships cannons and powder rooms called the magazine, or armory. A lower deck found below all others was one that was necessary to contain the ships ballast. The ballast usually consisted of hundreds of manageable rocks or iron pieces that were needed to stabilize the ships tilt or lean under sail. Cargo space was somewhat purposefully limited aboard pirate ships by the necessity of not overweighing the ship with non-essentials that would cause it not to be as fast or maneuverable when in combat with other ships.

The largest ship Kidd ever commanded was the captured *Quedagh Merchant* taken in 1698 in the East Indies which later sailed to North

America, a very large vessel rated at about 400 tons. The ship was renamed the *Adventure Prize* that was later set on fire and sunk in the Caribbean. All accounts of ship sightings around Nova Scotia do not indicate the *Adventure Prize* was ever seen that far north. There is however a pirate tale of Kidd possibly having interest in Oak Island resulting from a story told by an unknown person that gave him a map with an X marking the spot of hidden treasure on an island in Nova Scotia. It is not known whether or not Kidd ever made it to Oak Island on board any ship! On May 23, 1701 Kidd was hanged for the crime of piracy and murder. If Kidd did go to Oak Island it would have to be before that date, and that is hard to verify when tracing his history as recorded history has him being elsewhere.

Ships are rated on their cargo carrying capacity in tonnage weight more so than actual dimensions of size. The bigger the tonnage means the bigger the cargo. Most English ships that sailed during the era of piracy were usually no larger than 300 tons thus restricting their cargo holds to a fraction of an inconceivable load weighing two million pounds which converts into 1,000 tons of gold. There were however ships that may have been able to carry that much gold in one trip, Spanish merchant ships of the galleon or barque classes did have that larger capability.

Next there is the discovery of what has been described as the hair-like fiber of the coconut tree that was recovered during some of the excavations by the Lagina brothers. There are no coconut trees in North America. They grow in the tropical climate areas around the equatorial regions found worldwide. It is believed that coconut fiber was used as a filtration system to keep sand and dirt out of the tidal flooding booby trapped tunnel system found along the island's shoreline that leads to the "Money Pit." This find does indicate someone brought the fiber to the island with forethought about using it for that special construction need. The idea of using such a foreign ingredient in this construction was very well planned in that it involved a considerable amount of time needed to obtain those materials. I don't know how it would relate to the cross, except that it does imply there was a very high level of engineering used for both the cross and the flood tunnels. The thought process necessary in the preplanning and intellectual ability in understanding how to use hydrodynamics to engineer the construction of the flood tunnels, indicates a very unique individual or group during this era in time. Who, or whom could they be?

It is evident that someone, or some group of individuals did have very special knowledge of how fluids (water in this case) acted during the inflow and outflow of tides that could be controlled with the use of coconut fibers as a debris and sand filtration system. The person responsible for implementing the use of coconut fibers in this complex and very ingenious system is a very well educated and creative mastermind. I have a few thoughts who it might be that will be discussed later. The science involved in the use of coconut fiber had to be known as the fiber has the ability to swell to 15 times its normal size went wet, and is also free of bacteria from most fungal problems which causes rot! Coconut fiber has been used to make rope, mats for flooring and sleeping, and even woven into food containers that have been used for protecting packaged goods. The fiber from the coconut tree was also used to seal cracks or openings between a ships planking as it would swell to seal any gaps, an interesting fact. The regions of the earth most known for the growth of the coconut tree are: India, Java, Malaya, Sri Lanka, Philippines, Mexico and their surrounding tropical regions.

The first recorded use of the fiber was around the 3rd century BC and more so in the 11th and 13th centuries when it was used in construction by the people living in the areas around the Indian Ocean. The Europeans first used the fiber to chink leaks in their ships in the 1700s. Thinking about how the knowledge of the coconut fiber would help in solving what happened on Oak Island still leaves the questions of who and when. The coconut fibers were carbon dated to be between 600 to 800 years old, making the suspected time frame of their usage on the island around the 12th to 14th century, give or take a hundred years either way. That seems to be a much earlier time then when it was more frequently used by mariners sailing in the Atlantic. Other time indicators that have been used are based on analysis of the timbers that were used in connection with the finger drains uncovered at the beach site as they also fit within this time frame. The thing to glean from finding the fiber in the way it was used suggests that the designer of the projects on Oak Island had a very specific high priority preplanned mission! This was a very complex and strategically planned objective to hide something of what is yet unknown! Just more mystery!

Captain John Welling working as the head of operations for the Oak Island Treasure Company discovered a formation of stones in 1897 shaped like a triangle on the south side of the island. The stones were found about 50 feet inland above the high tide line of the island pointed in the direction of true north. Magnetic north would require a compass where true north relied upon the position of the Northern Star at night, an ancient old navigational reference used by sailors. The triangular formation was made by using numerous stones measuring about a foot in diameter that created an equilateral triangle measuring about 10 feet from the base to its apex. The triangle also appeared to have a bowed or arched bottom, connecting the two bottom sides of the triangle. There is a straight line of rocks laid to the left of the mid point on the arc bottom up through the triangle to its apex. The shape seems to represent a sextant! When Captain Welling hired a surveyor to investigate this curious formation, it was discovered the directional configuration led to some large boulders at distances measured in rods. The use of rods for the measurement of land distances have been used in England since the early 15th century, and was originally named as a perch. The rod is equivalent to the measurement that equals 16½ feet.

Unfortunately the configuration of the rocks forming the shape of a sextant were destroyed during some of the excavation work being done in later years by one of the searchers. However thanks to Mr. Fred Nolan there is a photograph of the stone formation before this possible clue was bulldozed away.

The next item found was the discovery of a small boulder engraved with the capital letter "G" that was framed inside an engraved square, one of the signs of the Freemasons, the legendary organization of craftsmen who originally began as indentured stone masons. The capital letter "G" is said by some to represent God, but mostly referenced to mean geometry in the Masonic culture of the Freemasons. The boulder appeared to be about the size of the bottom of an upside down wheelbarrow, and was discovered in 1969 by one of the land owners named Dan Blankenship. Blankenship had been involved with the island beginning in 1965 as an investor, and later became a land owner and a major treasure searcher on the island for over 50 years.

The Freemasons had their beginnings back in the era of the late Medieval ages between the 11th and 15th centuries. During that era of time is when people known as serfs or servants were tasked by their kings and other royals into building stone castles and churches in England, or elsewhere in Europe. Stone structures were more secure lasting longer than the ordinary wooden buildings that the citizens known as commoners used as housing. Stone structures, because of their longevity, made them a more valuable asset which could only be afforded by the royals and the churches. As the populations of the European countries grew, the craft of masonry would become more in demand. Beginning mostly in the 11th century, there was a need for the best craftsmen to travel to where the most need and money was offered. A king, or royal of some stature, would hire out his skilled subjects to construct castles, or fortifications for other royals and churches, then keeping the profits for their kingdom. Many times these building projects could take years to complete, sometimes requiring travel by the subjects to places far from home. As the Freemasons would began to travel farther away from home there was a need for them to be sheltered together, or at least be able to meet at a central location near the site being constructed. Usually there would be several masons along with their helpers needing a place to coordinate their activities. They soon would establish what were to be known as lodges. A lodge was just a simple gathering house or shelter that could be used by the masons to plan their activities.

Eventually these skilled subjects would either earn their freedom through good service, or just not return to their original area thus becoming somewhat of an escapee. Seeing that their need and skill warranted a special recognition above that of a mere servant to a land owner or royal, they eventually would form what is known as a guild, a type of labor union or tradesmen society. Now the group having formed a guild would soon give the masons the recognition to become known as the society of Freemasons. With the gaining of freedom from servitude, the masons could now operate freely for a fee paid by their royal employers or church leaders. One of the very first mason's guilds to be established was in Scotland during the reign of King Malcom III around the year of 1057. The Scottish guild was titled "The Masons' Company of Glasgow." In England, the Freemasons would become controlled by the "Mason's Livery Company of London" founded in 1220.

The formal recognition of the Freemasons would occur in the year of 1717 when the Grand Lodge of London was established. The Grand Lodge was like the headquarters for all of the smaller groups of masons coming from their local lodges to meet for major decisions important to the organization. It was during this occasion that the first Grand Master Mason was elected to head the organization. Ever since 1717 all Grand Master Masons have been required to be of noble heritage, requiring their successor to be appointed by the retiring Grand Master. Lodges would soon begin to form in France, Spain, Germany, even in North America around 1730 in the cities of Philadelphia, and Boston. As the colonization of the new continent of North America began to grow, states like New York, New Jersey, and Pennsylvania would also come under the control of a Provincial Grand Master Mason. A Freemasons' lodge was also formed in Florence, Italy in 1733. By the year of 1735 the number of lodges in England had grown from four to 126! A lodge would begin in Nova Scotia by 1738.

With the fact that the Freemasons were already an established guild in Scotland prior to the First Crusade in 1098, it is very conceivable to state that some of their members probably accompanied the Knights Templar to Jerusalem where they would quarter in or near Solomon's Temple. Both organizations had many similar principles and requirements.

The initial requirement to become a Freemason was of course the need to be a mason with some level of skill in the trade. They were the journeymen who were new to the trade. These were the laborers with less skill, and those with the most skill, were called the master masons. There was also a need for the person to be of irrefutable character with a strong belief in one God and a duty to obey His word. Similar to the Knights Templar organization there would be no women allowed to join. Freemasonry was begun on the premises not to be prejudice toward people, but to be open for admittance to all mature able bodied people of any race, color, or religion. But during the Middle Ages, women, or physically challenged persons were perceived to be less able to perform the tasks required to do the skilled work, and were therefore deemed unsuitable to be accepted into the guild. Eventually people of all the trades, or professional backgrounds would be admitted to join their ranks provided they met the new standards set by the Freemasons. There is an initiation process that all new members must submit to which has been scrutinized by outsiders as being secretive and ritual like, maybe even demonic because of the secrecy in which they are performed. The accusations directed at the Freemasons and the Templar Knights had come from those with their own agenda, who would attempt to destroy any group or society that has different ideals, or practices differing from their own. As history has proven neither the criticisms, nor the condemnations toward either group have ever been proven true, with most of those allegations coming from one source, the Catholic Church.

As the decades passed, turning into centuries, the Freemasons would eventually craft their society into becoming one of gentlemen and educated persons of importance. There was still the requirement that all had to perform their duties along side their brothers of other highly skilled craftsmen in many trades, who shared the same deep convictions aimed at becoming their best. They had in their beginnings of the organization adopted the religious doctrines of the Catholic Church with a strong belief in their duty to God. They also lived by a set of high standards required in the performance of their daily lives and crafts. In the early years of the organization duty was owed to the king and his laws that must always be obeyed, as well as compliance to the wishes of the employer. The Freemasons would structure their organization similar to many others with a ranking system based on what can be described as degrees of importance with each degree a step higher in importance. At the top of the hierarchy would be the Grand Master Mason with Master Masons and others lower in ranking. Obedience to the rankings and the degrees were to be earned while gaining respect. There is some belief that toward the end of the 11th century when the First Crusade began in the Holy Lands, some of the people accompanying the Knights Templar were also masons. Stone structures, like defensive fortifications and the rebuilding of castles and temples in Jerusalem, would require the need for skilled craftsmen to provide the guidance along with the labor necessary for the construction of those types of structures.

It is very probable that a group of masons were involved in the rebuilding of Solomon's Temple, as the temple had suffered much damage from previous invasions that had occurred over time in Jerusalem. During this time in history beginning with the First Crusade, when the rise of the Knights Templar occurs, coincidently the activity of the Freemasons is also very evident. There were others known as gentlemen, and the lesser knights that would come to join both groups. They all had united for a mission to recapture and to rebuild Jerusalem, believing they each had a commitment to Christianity with the belief in one God. Some came to the Holy Lands to fight the Islamic enemy, while others came to offer their labor in an effort to build a safe environment for the life of Christians in Christendom. Some came for the need of penitence for past sins, but some came as an act of chivalry who would die as martyrs in the eyes of the church. These two groups had a mission very similar in design; it was to create a new world defined by the construction of forts, castles, churches, and a culture safe for the growth of Christianity!

The construction of the various stone structures that would take place not only in the Holy Lands, but throughout Europe, was an enormous undertaking and a superb achievement. It has been estimated that thousands of stone castles, forts, sentry towers, and churches were constructed in countries reaching from the far southern lands of Israel to the northern castles of Sweden, to the eastern edges of Europe into the mountains of the Alps and Pyrenees of France and Spain. Portugal was a safe haven for activity of the Templars who enjoyed some autonomy after the years of 1307 as they operated under the guise of the Order of Christ. The country of France alone had more than 500 stone churches built in the first 300 years after the inception of the Freemasons. The most easterly European outpost or castle that was found in Serednie, Ukraine, near the border of southern Poland. Every Christian-occupied site in the

Holy Lands had some type of stone structure. Many of those structures covered large areas of land such as the Templar fort at Jacob's Ford on the Jordan River, known as the Chastellet. The Chastellet fort had outer limestone walls that stood as high as 30 feet tall! The whole fort was so large it would require as many as 1500 workers to complete before it was eventually destroyed by the Muslim leader Saladin. The number of commanderies, or Templar forts/castles alone were numbered to be as high as 50 at one time that could be found throughout the Israeli landscape.

Together, the Freemasons and the Templars were known to be associated with the construction of over 100 or more castles, plus other structures just in the Middle East alone. In Europe the number was even greater if churches and other fortifications are counted. The activities of the Freemasons in the construction of so many structures is still evident today with so many of the original castles, churches, and forts still in existence. They are a tourist delight when visiting the lands they occupy, with most offering an awe inspiring dramatic visual display of majesty.

The famous Rosslyn Chapel in Scotland, constructed in the 15th century, stands today as one of the most impressive architectural masterpieces ever constructed! This legendary church was founded by William Sinclair, who dedicated the chapel to the Knights Templar. The St. Clairs, also sometimes referred to as the Sinclair family, were Grand Master Masons. A coincidence? Probably not! Rosslyn Chapel was more recently noticed in 2006 with the debut of the movie titled *The Davinci Code* starring Tom Hanks. The book and the movie, both titled the same, are very controversial in their topic. The main subject here though is about the construction of the chapel with its connection in history as it pertains to both the Freemasons and the Knights Templar. The carved stone chapel is approximately 69 feet in length and 42 feet in height. The interior of the chapel is inundated with stone carvings portraying religious images along with the display of a carving done of a Knights Templar. A very interesting carving is the one that surrounds a very large window that portrays shapes appearing to be maize or ears of corn. A strange depiction of something that could only be found in North America! Maize, otherwise known as corn, was the staple of the indigenous people later to be known as the North American Indian! The maize engravings have been a point of much controversy as there has been much discussion as to how the knowledge of something so foreign at the time of the chapel construction could also be known in this far away land of Scotland! The carvings suggest, as some have declared, that either the Knights Templar or the Freemasons had been to America before the arrival of Christopher Columbus in 1492! That suspicion is well warranted in that the builders had displayed this unknown knowledge when the construction of Rosslyn Chapel was started in September of 1456! That is 36 years before the arrival of Columbus to North America!

There is also the upside-down carving of Lucifer bound by ropes that is found in Rosslyn Chapel. The figure of Lucifer shown in that position would imply he is the bound enemy of God being sent downward into Hell. The figure in that position has been equated with the rites performed in Freemasonry!

It is almost inconceivable to believe that these two groups with so many similarities were not associates in their past endeavors. The Knights Templar needed a commandery built for their personal housing along with stables for their horses. A commandery was used as a defensive fort like an outpost and as a tactical training headquarters. The Knights also required the support of entourages consisting of priests, cooks, makers of armor, clothiers, metal smiths, and other ranks of soldiers that needed living and working quarters. All structures had to be reinforced to prevent an enemy from breaching the interiors that could leave the occupants vulnerable to harm or death. Stone constructions would offer the best protection outlasting any other material known for the time. The Knights would train in the commanderies located throughout Europe for eventual service in the Holy Lands. The Freemasons were definitely a needed resource of skilled labor that could complete the mission of Templar expansion along with the rebuilding of any structures wherever necessary. The Freemasons could also be trusted to keep building secrets where escape passages or hidden vaults had been incorporated in the design of the forts and castle like commanderies. Secret passageways would allow non-combatants a safe route for escape during enemy attacks. Other secret hiding places would be used to protect valuables or documents. In an earlier chapter, there was a mention of one of the battles of the Knights Templar that took place at their fort at Acre on the Mediterranean. That fort was known to have had many underground escape tunnels and passageways. The Freemasons and the Templars had similar convictions toward religious beliefs and obedience to a higher code of conduct. A full and complete historical description about the Freemasons, their values, and their contributions to history can be found in a very good book written by Jasper Ridley, titled *The Freemasons*. The book is well written, and describes the Freemasons' society in all of its trials, tribulations, accusations, and truths.

Back to the Freemasons and their works. These two groups, the Templars and the Freemasons, had their ups and downs in history dealing with contentious kings, popes, and politicians. The Knights would, however, succumb to a hostile environment in the 14th century and would almost be completely abolished on eventual false charges of heresy. The Freemasons are still in existence today, and through the centuries have had some very prestigious members. The Freemasons' own belief was that God was the very first mason in his creation of earth and all of life. He was the master architect of all! Some of the Freemasons' members that have been known throughout history include people like Leonardo Da Vinci, Mozart,

Franz Hayden, Voltaire, Ben Franklin, Paul Revere, Franklin D. Roosevelt, and many others including signers of the Declaration of Independence, and a third of the signers of the U.S. Constitution. The list of well known and historically famous people is quite interesting and much lengthier than what is written here.

The question as to who did the engraving of the Masonic letter "G" on a small boulder framed inside of an engraved square found on Oak Island leads one to believe it must certainly have been done by a Freemason. The stone was not found until Dan Blankenship discovered it in 1969. The real question then is by whom and when was it engraved? Maybe even why? The dating of the stone to my knowledge has not been verified. If those questions could be answered, then the answer to what is the treasure on Oak Island could stand a better chance of possibly being solved. However, since more than one person who has been involved in the searches performed on Oak Island after the discovery back in 1795, have been masons, opens the possibility of it being done after 1795 but before 1969! However, the letter "G" was not used in earlier times as may be thought. The letter "G" only became a part of the Masonic symbol after 1730 when it was first introduced in America at the Philadelphia lodge by one of it's founders, Benjamin Franklin! What this means is that another Freemason could have done the "G" engraving prior to 1795, but was not one of those known to be an Oak Island searcher after 1795. It also means that any Freemason prior to 1730 did not know to use the "G" letter symbol, which rules out anyone that may have been an earlier suspect. So the stone was engraved by someone unknown as a searcher after the years of 1730, but before 1795? Who could that be?

Frederick Blair, a Freemason began his involvement on Oak Island in 1893 when he established the Oak Island Treasure Company. Another Freemason, Gilbert Hedden, becomes involved in 1935, and in the early 1950s, Mel Chappell, a Grand Master

Mason of Nova Scotia, becomes a partner. Mel Chappell's father William Chappell, was one of the earlier investors in Oak Island. Any of those individuals could have been the mason involved in the engraving of the letter "G" on the small boulder. The letter "G" was surely left by one of the treasure seekers as a permanent notice to others that a mason had been on that island. There is no known documentation though proving who may have done the engraving! As of 2022, there are no living witnesses to ask about the engraving, or whoever could acknowledge anyone knowing of the carver. Was it done by any of those Freemasons mentioned to be active on the island after the year 1795 or was it done prior? Masons do leave signs where they have worked. The question of who did it still remains another mystery!

Other Masonic symbols or signs include the drawing of the Acacia tree which is found in Israel. The significance of the Acacia tree is that it was used to build the Ark of the Covenant that housed the Ten Commandments. It was also the wood used in the construction of the traveling tabernacle, the movable temple for worship used by the believers in God. That symbol of the Acacia shows the reverence toward the belief in God. The item most used as a symbol to identify the Freemasons is the carpenter's square, a right-angled, "L" shaped tool that is overlaid with a mechanical drawing compass.

The capital letter "G" also stands for the power of a Gnostic person as having a special knowledge about spiritual things learned through devout faith in one true God. The believers that were considered to be Gnostics during certain times in history were considered not aligned with the beliefs of the Catholic Church and were condemned as being heretics. At one time in the early years of the Freemasons, the Pope actually ex-communicated the organization as also being heretics, probably because of their practice to hold secret meetings and rituals. Any time a group or individuals become protective of their doings out of view from non-members, it seems

Freemason Symbol

to never fail that they become subjects of ridicule and unfound accusations of being evil in some way or another.

The five-pointed star, otherwise known as a pentagram, symbolizes many things and is used by the Freemasons to identify their Five Points of Fellowship. The pentagram has been found throughout history to symbolize things with a special significance. The star was found on many buildings in Jerusalem during the time of Jesus to identify Christian houses and places of worship. It was frequently used on the pottery of Jews to indicate where it was made, like a trademark or ownership stamp. The pentagram is one of the earliest symbols used in Christianity starting with it being thought to represent the five wounds of Christ, his pierced hands and feet, plus the wound from a spear's thrust from a soldier to his side. Pythagoras, a great philosopher and mathematician, who gave us the theory in geometry of a right triangle that states when side A is squared plus side B is squared, they equal the square of the hypotenuse. He also described the symbol of the pentagram as being the one shape that is of perfection.

It is important to understand the use of the pentagram as the Freemasons' five-points for their foundation of belief and commitment to their brotherhood, as being those elements that aim toward perfection. The first point is for fellowship toward the needs of others and especially those in the brotherhood of the Freemasons. Second is the belief in the power of prayer that can help bring good when belief is strong through commitment to God's will and way. Third is one of trust that needs to be earned with commitment to secrecy of those words shared in the brotherhood as being guarded as sacred. Fourth is to support those in the brotherhood by not harming them with disrespect or with words of harm to their character. Fifth is the point for the sharing of wise counsel to others in need. There will always come a time when good support is needed to help others through some difficulty, and a Freemason needs to follow the ways of the Grand Architect, God, who would be there to give aid.

The Freemasons' goals have been simple: make good men better in all they do while developing moral, spiritual and intellectual growth while striving to be a good example to others. Freemasonry is a fraternal organization that still exists today throughout the world.

The figure of Tanit found on Oak Island basically resembles the shape of a small Latin cross with the very top of the vertical shaft having a rounded shape similar to the appearance of a human head tilted to one side. The head has a small square like opening in the center, and the vertical bar below the horizontal cross bar has a slight divide possibly indicating two legs. The horizontal cross bars on each side of the vertical shaft are to represent outstretched arms. Tanit is an ancient goddess symbol with origins from the area of Lebanon located on the eastern shores of the Mediterranean, north of Jerusalem, that dates back to 800 BC. The symbol was used as a representation of the Phoenician moon goddess named Tanit beginning around the 5th century BC after the Phoenicians left the city of Tyre, fleeing to Carthage in North Africa where Tanit became the

Lady of Carthage. The historical depiction of the ancient Phoenician goddess Tanit symbol is found with slight differences than the one found on Oak Island. One Tanit-style figure has a large, open circular head with what appears to be a triangular shape representing a woman's skirt below outstretched horizontal arms. The image was created to be feminine on purpose as a representation of a goddess. There have been several items of worship that were feminine in the pre-Christian era. The small differences in the shape of Tanit depend upon what culture and period of time she was worshiped. Another slight variation is shown on a statue of an Egyptian pharaoh holding a Tanit symbol in each hand with the head at the top as being very circular with the center void. All symbols of Tanit do however represent a feminine goddess with a likeness to the shape of a cross or that of a woman with outstretched arms.

The one discovered on Oak Island is very likely a slight modification of the original shape, but is one that surely has a very similar meaning and that is of a female goddess. For a Templar to use an exact replica of the very early Egyptian or Phoenician Tanit would be considered heresy. I believe a possible scenario is that a Templar, or a person having acquaintances with the Templar Knights, introduced them to the historical worship of the Goddess Tanit while in the Holy Lands. The Knights learned about the ancient cultures' deep respect for and worship of Tanit that was practiced in the region and equated it to being similar to the Templar respect for the Holy Virgin Mary. Being devout Christians, some may have decided to fashion their own Tanit as a way of worshiping the Virgin Mary by creating a similar shape likened to the Latin cross. The purpose may have been twofold: one that is of the worship to the Virgin Mary and the other to be accepted by their possible Egyptian and Syrian captors as a saving symbol that reflected the local ancient Tanit culture. This is only speculation, but if you were in enemy territory, the best disguise to avoid capture or death is to appear to be as one of your foe. I know that Knights of the Temple were known to fight to the death when facing their enemies, and I would not want to insinuate otherwise. But not all Knights died in combat! Some came to the Holy Lands for brief periods of time before returning back to their homelands to expand the Order. The Tanit symbol may have been, as we often say in today's world, a keepsake from the Holy Land.

The adoption of the Tanit symbol may have been another secret that the Knights Templar kept from the prying eyes of the Pope and especially from the King of France as they were already being accused of heresy. No need to advertise for another accusation of heresy that could lead to condemnation. The Knights held their secrets well as it was part of who they were. It is still not known if they found something of immensely historical and religious worldly importance under Solomon's Temple. Some have suspected there may have been a secret that the Pope and the Catholic Church wanted so badly to possess that it was the real reason for the Knights eventual demise!

The use of symbolic shapes have been used as a form of worship since the beginning of humankind. The ancient Mediterranean area produced many different named shapes that were worshiped with most referred to as being feminine, or goddesslike. Tanit was worshiped as being akin to the Egyptian goddess of war, Baal Hammon, who was thought to be a female virgin from heaven. Another goddess of worship from this Mediterranean area is one of the Egyptian Goddess Isis. Isis, in Egyptian culture, was considered the mother of all pharaohs, a creator, a preserver of life, and was the queen to the highest throne. The story about Isis is quite interesting in that it describes the life of a mythical queen whose husband is abducted, murdered, and died on a tree, and fearing for her life, she goes into hiding. Her eventual goal is to retrieve her husband. She miraculously becomes a virgin mother giving birth to a son. Some similarities here to the story of Jesus and the Virgin Mary. Isis was worshipped until 529 CE when Christianity overcame idol worship in most areas.

The worship of an earlier Phoenician Goddess named Astarte, who was considered a mother figure, eventually waned as Tanit became more popular in assuming the role of Mother Goddess. After the Romans entered the scene around the Holy Lands where they learned about Tanit, they equated her to their Goddess Juno Lucina, mother of childbirth. Tanit was referred to by the Romans as "The Heavenly Virgin" who had control over the sun, stars, and moon. The Romans eventually destroyed Carthage in one of their many invasions into the southern Mediterranean lands, but respected Tanit so much that they erected a statue of her in Rome with the new name of Caelestis. The symbol of Tanit is found on gravestones in the regions of the Holy Lands. Tanit can also be found on many temples in the north African regions of modern day Tunis that was once known as Carthage where the Phoenicians had lived. If you take a look at an image of the ancient Egyptian Goddess Ma'at ruler of truth, justice, and balance, she is shown holding a symbol of Tanit as if in admiration. The influence of Tanit as a Goddess has been worshiped not only in the Mediterrean areas of Carthage, Egypt, and Rome, but also as far away as Spain, even the isles of Malta and Sardinia.

Assuming that all of the Tanit shapes have the same meaning as being a Goddess, as they definitely appear to represent the same concept, it is very conceivable that the worship of each is also shared with the same meanings as their followers. According to historical evidence about some members of the Knights Templar being jailed in Domme France along with evidence of a Tanit engraving found on the wall of the prison supposedly done by the Knights Templar, it is not without logic to extrapolate that the Tanit found on Oak Island has a Knights Templar connection! Should this be true, then there is reasonable belief that there was a Templar presence on Oak Island sometime in the past!

Knowing that the Knights were a religious organization committed to protecting the Holy Lands including safeguarding any religious artifacts they may have found would give credence to the idea they may have concealed something of importance on the island! Remember that the Knights Templar have claimed to be the guardians of the Holy Grail and the Ark of the Covenant! With the possibility of the Knights Templar having been present on Oak Island and being guardians of sacred religious items, could any of those items be buried on the island?

If the Templar are the true architects of the events on Oak Island and they did bury something, then it is highly probable that some type of a sign would be left to indicate that the island is one that has an extremely religious existence! The reason for that assumption is in the evidence that is always found at any Christian hallowed site, the symbol of a cross!

The symbol of the Goddess Tanit and the cross share a secret!

Nolan's Cross

The Oak Island story has become a maze of mysteries with each one uncovered seemingly more intriguing than the previous one, yet not a single one of these mysteries has come close to being solved. We now know the story that began a couple of hundred years ago when a young man by the name of Daniel McGinnis found a suspicious area of sunken ground on the island after investigating some suspicious activity that he had observed one evening. What he found could have been the trace of some old pirate activity where a treasure may have been buried or some other individuals doing something secretly. There were too many sketchy tales about pirates roaming the waters up and down the east coast of North America and Nova Scotia to think some might be really true! There were at least a few suspects though that could be named. If it was treasure buried there beneath the depression in the ground, why Oak Island? When was it done? Who did it? How can it be retrieved? Yet the biggest question or mystery of all: what is it that may be hidden there, if anything?

When you begin to think about some of the things that have already been found on Oak Island, there is one discovery that seems to be very perplexing which comes to my mind. What is the possible relationship between the supposed pirate's treasure buried there and the discovery of some large boulders configured into the shape of a very large Latin style cross that spans hundreds of feet? At first thought, these two things would seem to be completely unrelated. There couldn't be any relevance between the two of them at all or could it? The cross symbolizes religion and love while the Pirate treasure theory infers to the ill-gotten booty acquired by means of nefarious plunder. Maybe there could be a connection! Was it possible that Fred Nolan, the surveyor who mapped all of Oak Island and discovered the cross back in the 1960s, had found something significant that others had overlooked? Yes! He was a smart man who would eventually spend many years trying to solve the purpose of what he had found and why it was there on Oak Island. Is it possible there is a hidden connection between the legendary Money Pit and Nolan's Cross? Why are they on the same island? Has there been some trickery done here with one or the other being a distraction as some have said has been done? Some have proclaimed that one is only meant to be a diversion away from the real treasure. The Pit and the Cross are each very different and uniquely unrelated. Or are they?

This book will try to answer a most important question that will hopefully validate the importance of what Fred Nolan disclosed in 1981 resulting from his survey. Although he never completely described what he thought the cross meant or why it was there, Nolan's reasons for not saying what he thought could have been to avoid accusations from others of him being too imaginative or worse. Nolan was a man who dealt with facts in his profession with his word needing high respect. I am sure he did not want to undermine his reputation by saying something that others would doubt, or not understand. Yet, he may have located the most important clue ever found on Oak Island that could explain its deep secret! Through my own curious research of analyzing what could be considered applicable to Oak Island, I have stumbled onto what the true purpose was for putting the cross on Oak Island! I will reveal to you what it is all about and how I came to that conclusion.

It could be the most important historical find ever discovered, not only on Oak Island, but even in North America!

It is unfortunate that Fred Nolan will never know fully what he found on the island as he passed away on June 5, 2016, just one month before his 90th birthday. Mr. Nolan did though mention to Willam Crooker, a one-time confidant and author of the book titled *Oak Island Gold*, that there was more to the Cross that was yet to be known. Nolan had developed his own ideas from some of the things he had found on the island during his more than 50 years of owning a piece of the island. He also kept track of what others had discovered in their many attempts at trying to solve the puzzle of the Money Pit. Such a mystery this all has become but there is an answer hidden somewhere about those who created the mystery. If they did leave any clue as to who they were, what could it be? Whatever the solutions may be to the mysteries of Oak Island, whenever they are resolved, it is evident that they will not be solved through a mere cursory analysis! These are entangled mysteries that run deep with different events in history, but all had a similar goal!

Earlier in this book it was explained how Fred Nolan got involved on Oak Island. But basically it was because of reading the book written by R.V. Harris *The Oak Island Mystery* published in 1958 which chronicled the history of events that surrounded Oak Island. A well written story!

A little refresher here about Mr. Nolan: he was born in Halifax, Nova Scotia and became one of the first land surveyors around Halifax to become involved in mapping out the many local community areas during his tenure. Being a land surveyor required him to do much research of historical records and documents issued by governments, or others relating to ownership and title to property. Another aspect of being a land surveyor is the need to be extremely proficient in mathematical skills in the use of geometry, distance calculations, area measurements, trigonometry, and understanding latitude and longitude degrees that relate to locations on the earth surface. The need for all of that knowledge and more, enabled him to correctly survey and create plat drawings for legal recordation of lands. So it was very understandable that one of the first things Fred Nolan was interested in doing on Oak Island was to map its layout. Living only a few miles just to the north of Oak Island in Halifax, Nolan was determined to venture to the island to seek permission for doing a survey from the then-owner of most of the island, Mel Chappell. There had been, as written about earlier, some survey work that had been done by others in 1936 and again in 1762 and 1765. Nolan wanted to do his own survey so he could actually explore the islands entire 144 acres. It took Fred Nolan almost two years to do the much involved process of laying out his many survey lines that often required him to cut through acres of vegetative growth in order to place his survey markers before finally being able to map all he had surveyed. All of this had taken him away from doing his normal surveying jobs for paying customers. His searching would seem to become more important than his livelihood. You could say he had become obsessed with finding out what was really on Oak Island!

After Fred Nolan had bought the land around the swamp, and things had settled down between he and the other owners, Nolan would find some interesting evidence of unusual past activities on the island.

The things that Nolan found on the island in his pursuit to learn what the island was began with the discovery of a small rock with the carving of the letter "G" appearing inside a rectangular box that was oriented vertically. The letter "G" was found on many buildings throughout the history of Freemasonry which was formally organized in early England around the year 1717. This was all described in earlier chapters, so no need to reiterate it all here.

Another item of interest found on the island was in the shape of a large equilateral triangle that had been formed using 12- to 14-inch, round beach stones. The triangle had been constructed with sides measuring about nine feet in length with a base of about six feet and the shape appearing to be pointing due north. William Chappell and his head of operations, Captain Welling, had found the stone configuration in 1897 while both men were members of the Oak Island Treasure Company. The triangle was discovered to actually be pointing in the direction of the Money Pit. It seems probable that the large stone triangle had been made as a directional reference placed by whomever had created the Money Pit. The triangle had been placed about fifty feet inland above the islands high tide line and could have been found by approaching sailors in a boat looking for the rocks as a direction sign that would lead them to the Money Pit. When the stone triangle was found back in 1897 it was thought to have been left by a previous search group and was thought not to be of any significance. Nolan fortunately had taken a photograph of the rock triangle in the 1960s along with some notations, measurements, and distances for his records. In the picture that Nolan had taken of the rocks, they seem to have been located in their same position over a long period of time, evidenced by signs of having settled in the ground while the area around them seemed to have never been disturbed. A good thing he had taken some photos of the area because another searcher, who later was excavating around the Money Pit area, destroyed all physical trace of the triangle after doing some bulldozing work.

Fred Nolan also found more items of interest around his favorite spot called the swamp which was at the southeast end of his property. As a surveyor, Nolan was always looking for anything that others would have left as markers in the past. That was part of a surveyor's task when checking properties. He did find a few spruce poles about six feet in length that had been sharpened at one end. These types of poles had obviously been driven into the ground to be used as markers in identifying their location as an important reference point. But who had left them was questionable as there were no records of it ever being done. They didn't have any markings for identification and the spruce poles seemed to have been very old due to their darkened appearance. The darkening and almost blackening of wood is caused by weathering that occurs over a long period of time. There had been others from past groups searching the island that may have left the spruce markers, but what were they meant to identify? Were these the markers used to lay out a trail or road, or maybe they were used as a sight line that would lead to something of importance? Spruce trees are native to Nova Scotia and the Canadian regions, so their use in the area of Oak Island would not be out of the ordinary. Had the markers been of a different species, they may have been a clue leading to the identity of some unknown foreign searchers. Also without the availability of any modern carbon dating techniques during the time period the poles were discovered, they were of no help in determining their age which could give a clue as to when they were used or by whom. Yet another mystery!

The most important thing Fred Nolan ever discovered on Oak Island was something that resulted from his very meticulous survey work, and that was when he recorded the positioning of the six very large sandstone boulders. The boulders measured about nine feet in circumference with each estimated as weighing in the tons that were found placed hundreds of feet apart in the very precise formation of a Latin styled cross!

The boulders being placed at such distances apart would only have been noticeable by the efforts of a very observant surveyor after mapping their location with the realization that their placement could only have been designed on purpose by humans! Although the island had been previously mapped by others, they evidently had ignored the significance of these large boulders, or they just did not have the intuition to literally connect the dots! It was however a familiar sight to see boulders on many of the islands around Mahone

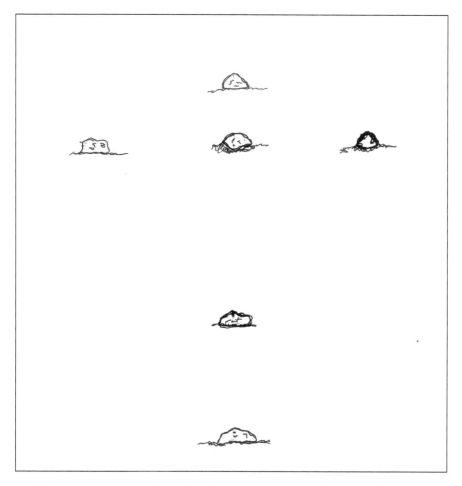

Placement of six boulders of Nolan's Cross to scale drawn by the author

Bay. Other surveyors may have just taken them for granted, but without their recognition by Nolan, those boulders found on Oak Island would probably still only be seen as unimportant!

So the very important question arises: why is there such a large configuration of a cross laid out over such an expansive area of land in the way it was it done? By whom, when, and for what reason?

First, one must consider that the formation of the boulders representing a cross is the most recognized religious symbol of Christianity in the world! That the design of a cross was surely meant to indicate that there must be some special religious connection intended for its placement there on Oak Island. What intent could that be? Was it meant to attract Christians who would be traveling to the waters of Mahone Bay? Even as large as the cross is, it would literally be hard to recognize by a casual sailor just passing by. Just think of all of the previous treasure seekers who have trod on the island and have not recognized any meaning to the location of those boulders. Most all of the attention on the island has been directed toward the Money Pit and the unwavering speculation of a treasure being buried deep in an underground shaft. The cross has all but been ignored by the bevy of searchers who have trampled over the island's grounds for over 200 years! The task of placing those huge boulders in such a precise alignment at over such an exact measurable distance was no easy task, even with the use of today's modern machinery! Remembering that the construction of this cross was definitely done before Oak Island was recognized in 1795, there is no record of any group participating in such a construction after that date. It is more than likely it was done decades before that time period or maybe even earlier!

How could any undertaking that would attempt to move such huge boulders, each weighing tons, that were placed over hundreds of feet apart into very precise locations, be accomplished without an enormous work force? That type of construction, without any doubt, would have required the need for dozens or maybe even hundreds of men along with the help from oxen or horses, plus moorage for ships near the shoreline that could offer transportation or accommodations. This project, in reality, was no small scale activity that surely would have been seen by anyone frequenting nearby, yet not be witnessed? What group could have had the potential to perform such a huge undertaking on this relatively small island and never be identified? Again we must ask what country, what organization, what military, what clandestine group, or what possible church had the motivation to construct such a cross? What purpose was it to serve?

We know from history that the pyramids were built using stone slabs weighing in the tons, but we also know that those constructions were performed by hundreds of slaves requiring decades to complete. Yes, they used some ingenious engineering such as hoists, ramps, and rollers, but the effort to move the slabs still required enormous energy. Considering there was no modern day powered mechanical assistance available when the boulders for the cross were maneuvered into their respective postions, it must be assumed it was all done as had been done in the past with only brute strength and simple devices!

There could though have been enough manpower capability as has been suggested in other chapters when describing the possibility of the work being done by the crews aboard some military operations of the French or English in Nova Scotia. All have been scrutinized, but there is more needed to investigate. There is only one national military group that may have been capable to do the task of the pit that can merit any suspicion. But, I don't believe they had any motivation to do the cross! That one group was led by Sir William Phips an Englishman who actually located the sunken Spanish galleon *Nuestra Senora Concepcion*. The ship was still laden heavily with tons of silver as she lay broken in two parts on the ocean bottom of the Bahamas in 1687. If anybody had any found wealth to hide, Phips had a good reason to do so more than many others. He could have done the pit to avoid giving all of the treasure to his government, but there is a strong suspicion someone else did the flood tunnels to keep others away. In either event, if true, these other suspects had no religious motive to risk exposure in committing enough time and labor to engineer the site for the cross!

What makes Nolan's Cross so very baffling is that it is there and no one knows why.

Nolan's Cross - A Map?

When people are about to consider taking a trip somewhere, or want to locate a specific recorded place, they usually try to find a map to help determine a route to travel. They may be looking for a certain locality or a special place of interest. Maybe even getting ready for a vacation to go somewhere they have never before ventured. Maps have been used for centuries to guide humans to places they sought to travel to or were created to record where they had been. The ancient peoples of centuries ago were limited in their access to maps or navigation devices such as what modern technology has given us today, like a handheld GPS device. The earliest methods of navigation used by humans for travel to far away places was to use the heavenly stars at night, the daily rising and setting of the sun, or just the physical geographic things like a mountain, or a body of water. Maps do appear in many different forms. There are the topographic or physical kind which show the earths landscape and geographical features in two dimensions like mountains, rivers, deserts, and other natural formations. Then there are the more familiar modern day city, state, and world maps that are available to show us where something is found, enabling the user to plot a route of how to get to that destination. There are types of maps for just about anything imaginable. Some other recognizable ones include weather, time zone, zip code, telephone area codes, nautical maps, planet maps and more, all designed to easily guide us in gaining information about wherever we seek. Today the view of the world has even been captured in a collection of reference maps that we call an atlas. With the use of an accurate globe of the Earth (a form of map), you can visually travel quickly around the Earth.

Could Nolan's Cross be a map? If it directs someone to something or somewhere, or is used as a symbol indicating where something is found, then why not? For centuries, the cross has been recognized by Christians as their most important religious symbol ever since the crucifixion of Jesus. It has been accepted as a sign of Christianity and something of holy significance no matter how it is displayed. Maps usually incorporate some kind of symbols to indicate certain features of importance, like a fork or spoon to indicate a eatery location, or a small red cross to locate a hospital. Highways usually are shown to have a number inside a shield shape or a tombstone to show where a cemetery is found, and so on. If someone was creating a map that others could use to find something in a certain location, they might use some form of a symbol to show where that something was to be found.

Suppose you made a map to hide some valuables that you wanted to leave your heirs to discover after your demise. Let's say it's in your backyard, buried five feet deep being four steps east of the old oak tree in the direction of the rising sun. The map would in all probability show a tree in your yard, an image of a rising sun, and showing four human feet, a shovel with the number five on the handle, and a treasure chest to imply something valuable buried in that location. So the items used on the map are actually symbols that describe some things of importance that are necessary toward accomplishing the goal of finding their inheritance. If Nolan's Cross was drawn on a piece of paper, could there be some form of interpretation suggesting it is a map? Depends upon how it is viewed.

There have been a few attempts in recent years by a small number of theorists that have tried their techniques in manipulating things on the island in order to explain their version of Nolan's Cross as a possible treasure map. None though have been successful in directing the searchers toward finding anything of value, nor have they proven anything that would even give promise as to the cross being a directional tool! A couple of the theorists have suggested using Nolan's Cross like a compass that points to what they believed was the possible direction to the treasure site. Of course they all have come up with negative results every time. It just seems amazing that so many have tried so many different approaches in their attempts to finding the treasure by using what they thought were good ideas with honest intentions, but none have ever produced any legitimate leads to any booty. I guess everyone would like to solve a mystery; it's in the natural makeup of the human mind to be challenged intellectually to seek a solution for something that wants to be answered. However, there seems to always be something everyone just can't quite make sense of when they attempt to do something that maybe isn't intended to be done! Maybe the cross is not a map or a directional indicator pointing to what they had hoped – treasure! Only time will prove what the Cross really is; maybe it's just a symbol to show that Oak Island is to be recognized by Christians who were thankful in finding the land that would lead them to the new world some would call Acadia! The place of a new France!

There is still the question of what is the true purpose of the cross! If the cross is not a map, then maybe it is just a symbolic site marker for Christians. Could that be its only purpose? If it was just a symbolic site marker for Christians, then it must have a religious reason. What reason could that be, and why there on Oak Island?

It is time to finally uncloak Nolan's Cross!

The Cross Uncloaked!

The mystery of Oak Island has been protecting a unique secret very well for some time, but that time is ending! The tools to solve the mystery will not be with shovel and excavating machine, but with the mind that uses mathematics and geometry!

Any of the necessary evidence that is known or recorded about events that may involve Oak Island have been presented. The history of those who participated in the searches have been introduced sufficiently enough to garner an understanding of what needs to be known. The suspects named have only been rumored to be involved, or those who actually had some potential to be involved, have been determined as unlikely in actually accomplishing the deed! With no real evidence of any participation by any of those suspected, there is now a need to analyze what artifacts have been found during all of the searchers efforts which could possibly give some clue as to solving the mystery. The important artifacts that have been found over the decades have been scrutinized as well with each of those only adding more questions to the total mystery of who, what, why, and when! There are, in my opinion, after considering all that has been presented, only two important finds or artifacts left to further consider which could offer any possible indication as to what Oak Island is all about!

The first item is the lead cross that is described as being a Tanit that was discovered on the beach by Rick Lagina and Gary Drayton. That item alone has given the searchers – or should we say mystery solvers – an undeniable clue that leads straight to the Knights Templar! It has already been written about earlier how this artifact has a connection to the Templar travels, along with the fact that the lead material it is made of came from the eastern region of France during a time of known Templar occupancy! The History Channel TV series *Buried: Knights Templar and the Holy Grail* with Mikey Kay and Garth Baldwin that aired in 2018 showed Tanit etchings on the walls of the Domme Prison in France that once had been the holding cell for some Knights Templar. A very obvious connection!

The second, and probably the most important of all finds ever, is the expansive stone configuration known as Nolan's Cross. As stated earlier in this book, my suspicion arose after analyzing everything else about this island, and its mystery that is actually several mysteries combined into one! There had to be something about this cross being there on this particular island next to the legendary Money Pit! What a strange partnership; it was unlike anything ever found anywhere else on earth. With the realization that the placement of these boulders in forming a cross did not happen naturally, curious minds want to know how it was done. Once that question is mused for a bit, then the next curious inquiry would be to want to know why.

There had to be some special reason of such enormous importance to create both the Money Pit and the huge cross at the same location in such an isolated area. Why things have been done on Oak Island is by itself a big mystery! Why not some other island in North America? Why even an island? Could the same events have been done in Europe? Why not? So many questions to consider asking; that's why the mystery is so complex and mentally challenging!

How did it all come to this that there are only two things of real importance to focus on for a solution. First, what must be considered is that there has been no treasure found that could give a clue as to the type of treasurethat may be buried. Second, there is no written record of any event prior to 1795 that can truly explain "who" did Oak Island! However, it should be easy to surmise by now that what happened there was surely done in secret; it had to be some type of a covert operation! What group would have that kind of a commitment to make it their secret mission? The suspicion about what has occurred on Oak Island could have been accomplished by just one group if it was executed through the culmination of several events that led to one purpose! But only if all had been collaborated secretly through the decades to hide something so precious, it ultimately had to be concealed to protect it!

With the discovery of the huge cross on the island, it would be hopefully be obvious that it was placed there to refer to something that has a religious meaning as in Christianity! The question is: why is it there? It could be, as stated earlier, just to mark the island for Christian settlers coming to Acadia as being a safe place of refuge. But how would they have known that since the cross was never discovered until an observant land surveyor named Fred Nolan had mapped it in the 1980s!

Could this cross have been configured there to mark the island as a gravesite as is the usual tradition at a Christian burial site? What could be buried there that would cause such a commitment as to require this type of an undertaking? Should the grounds in the area of the cross be considered hallowed grounds?

The cross of boulders needs to be examined more closely to see if there is a hidden clue in its unusual size and configuration!

The first thing that needs to be done is to examine the design of the cross to see if there is any evidence of it being any different than that of a regularly seen Christian cross. The typical Christian cross does have some accepted dimensions that have been used over the centuries with it mostly adhering to that of a Latin styled cross. A typical Latin styled Cross used in most Christian displays is constructed in a proportional design of thirds as follows: the entire vertical shaft is 3/3 of any scale with a horizontal cross member of 2/3 the vertical total, and the vertical section above the intersection of the cross beam would be at 1/3 of the vertical with the lower half below the cross beam at 2/3. So for example, if there is a nine foot beam in total, then the horizontal cross beam would be located at three feet below the top with six feet below the intersection with the cross beam and the cross beam would total six feet across being centered with three feet on either side of the vertical beam.

Various styles of religious crosses

There are many different styled crosses used by different religions but none that resemble the Latin style cross so closely.

When we take a look at the measurements of the cross found on Oak Island, there is no thirds relationship!

The top of the vertical in Nolan's Cross is only 145 feet above the intersecting cross beam, which measures 360 feet on either side of the vertical. The top is less than ½ the distance of either side! For the cross to be a usual Latin style, the top would also have to be 360 feet above the vertical intersection! The combined measurements of the vertical that is below the horizontal beam totals 722 feet, which is close enough to the 2/3 distance of the proportional size normally found. However, there is a problem here due to the location of a sixth boulder being introduced to the overall design. There does not need to be two boulders found below the horizontal beam.

The one boulder located at just 429 feet below the horizontal beam only equates to about 1.2 times the length of either side of the horizontal beam of 360 feet. That boulder, if placed in the proportional position, should be at 720 feet! So why are there two boulders located in the vertical beam below the horizontal intersection, and why is the top vertical above the intersection of only 145 feet? This cross is designed in a special way to convey some form of a hidden message due to its non-similarity to a normally seen Latin style cross!

At first glance, after viewing the drawing that Fred Nolan had created of the boulder placements, it would seem to cause anyone to be curious about why there were so many boulders in the vertical position. Since the drawing of the boulder cross had now come to catch my interest, there became a demanding curiosity to examine why the cross appeared differently than what is normally displayed. Believe me: what was found was a surprise beyond anything that could have ever been imagined! The discovery was very puzzling to say the least, taking several months of trying to accept what was really hidden in the cross!

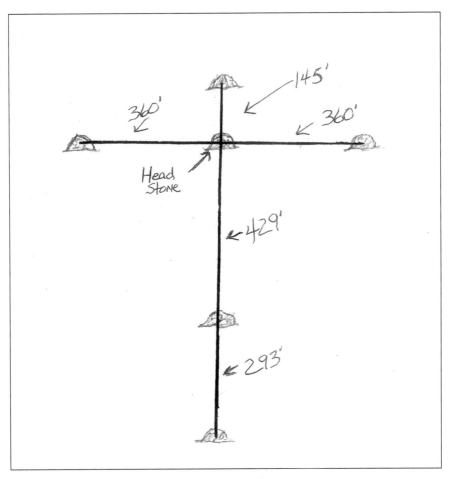

Nolan's Cross with original measurements
Illustration by the author

Let me explain.

As stated earlier in the book, I had been a mathematics teacher for several years working in the same school as my wife who also taught mathematics. My favorite subject when teaching was always the geometry lessons, especially when it came to spatial relationships of how shapes are integrated in our everyday world. That area of geometry also blended well with real estate in understanding home designs, land layouts, and other architectural items. Needless to say that all areas of mathematics helped when discovering what Nolan's

Cross was meant to tell! There was something not only about the size of the cross being spread out over a vast area, but in the measurements used in the placement of the boulders. Being a math teacher with real estate as a second career had combined a lot of different areas of knowledge, including some surveying experience. How this configuration of the huge boulders had been arranged with the mathematics that was used gave suspicion that there was a lot more to this than what was ever thought! Earlier in my life, I was in the military and I had been introduced to cryptography as a crypto operator. This prompted a lifelong interest into the art of coding messages. It was necessary to also utilize some of this knowledge and experience to understand how to actually begin to decipher what the cross was hiding!

In studying the measurements of the cross, knowing that the thirds proportion was not really adhered to, there had to be a reason for the architect of the cross to deviate from that norm. Maybe the measurements were not a mistake and what was done was intentional! This is not saying that Fred Nolan made any mistakes. Quite the opposite: his surveying skills are what thankfully discovered the cross. The big red flag was the 145 feet vertical above the horizontal beam, and the location of the sixth boulder at 429 feet. What I then began to do was to put pencil to paper adding the measurements, dividing them, multiplying them, and grouping them in sets. There is no need to explain the whole lengthy process that was taken, but only a brief explanation of the outcome will be stated here. That process, eventually after some days of trying different approaches is what helped to eventually see a unique mathematical combination of numbers that when divided began to reveal a special number that was familiar to mathematicians – the number 144!

The number 144 is one of the numbers found in the Fibonacci Sequence. If you are not familiar with Fibonacci, he was a brilliant Italian mathematician living in the 13th century who devised the theory of number sequencing that states "each number is the sum of the two preceding numbers" and so on for infinity. Here is a short example: 1+1=2, 1+2=3, 2+3=5, 3+5=8, 5+8=13, 8+13=21, 13+21=34, 21+34=55, 34+55=89, 55+89=144 and repeats the process for infinity. His number theory also discovered that if the first number in the sequence is divided into the second number in the sequence you get what has been termed to be a Golden Ratio of 1.6. This is an infinite quotient which means the answer continues on forever as the result of one whole number being divided by another whole number and is referred to as Phi. Here is an example of that: 144 divided by 89=1.6179775281. This ratio is considered to be the Golden Ratio or the Divine Proportion, as it is believed designed by the Creator of things on earth and in heaven. The Fibonacci Sequence and the Golden Ratio can be found throughout nature in things like a conch seashell that has the spiraling of the Fibonacci Sequence! In the world of art, artists will use the Divine Proportion in their paintings as a way to balance the relationships of objects that are more pleasing to the eyes of the viewer. This balance, or golden ratio of numbers, cannot be written as fractions. Enough math here; there will be more, though.

So how did the number 144 appear in the cross that Fred Nolan discovered? It didn't; it was disguised!

At first glance, the numbers in the cross seem to appear as having no special relationship in the mathematics used in its construction, however, as stated earlier, by manipulating the numbers to see if there was any combinations or sets that could be distinguished, there were!

Let's begin with adding all of the distances from the top to the bottom of the vertical line of the cross. The numbers to be added are: 145+429+293=867. Nothing seems to stand out except that the numbers add up to an odd number total. Maybe just add the bottom two numbers of 429 and 293 to get 722; now there is an even number divisible by two. There is something here because if adding the sides of the horizontal lines of each side measuring 360 + 360, together they equal 720 – another even number. The vertical distance of the bottom two numbers added together are close to the total of the two horizontal sides! At this discovery, there was a need to analyze what Fred Nolan had recorded as being as accurate as possible and that there were no errors in his measurements. What had to be acknowledged, though, was that these measurements were taken where Nolan found the huge boulders after they had been in place for probably decades, or maybe even centuries. If the boulders weighed in the tons as they are estimated to be, and any settling into the ground had occurred due to their massive weight, then that settling would be expected to cause them to shift from their original positioning. Another factor in the measurements taken is where the point of reference was initiated upon the huge boulders. With each boulder estimated to be about 9 feet in circumference and in height, finding a measuring point to sight the distance to the next boulder could be different for each boulder. Trying to find an exact spot on each boulder would require making a permanent mark to have a repeatable reference spot. Just moving your measurement starting point from one side of the boulder to the other side can change any of the measurements that Nolan had recorded. Anyone measuring the boulders today could easily calculate differences in those measurements. The measurements are not necessarily incorrect at all, but are recorded as to the placement of the measuring device where placed on the boulders during Nolan's survey.

Understanding that any of these measurements could be adjusted because of the previously mentioned factors, it was time to begin some adjusting on my own to find if there were any mathematical connections that could be applied. I had already decided that the Fibonacci sequence number of 144 had my attention; now it was time to put it to the test to see if it could work on the other measurements.

Let's take for granted that the top of the vertical line of the cross is really 144 feet and not 145 feet, a very slight adjustment of only one foot that could easily be made. That now matches the Fibonacci number of 144!

Next, by taking the first number in the bottom of the vertical line 429 to see if 144 would divide into it, we get 2.97 or almost 3 times. Not a good match until the 429 is changed to 432 which is 144 multiplied by the number 3! As stated before, a slight adjustment here just to see what could be an outcome. This seems to be leading to something so the same will be tried on the bottom of the vertical number.

Divide 293 by 144 to get 2.03 times and that quotient does not terminate as a whole even number, so take 144 multiplied by two to get 288, which is another small adjustment of just subtracting 5 feet from 293! If we take those two measurements and subtract the amount of 5 feet from 293 to get 288, and add the amount of 3 feet to the 429 to get 432, there is an adjustment of only 2 feet. But, if we also take the 145 of the top vertical and change it to 144 feet, then that is another subtraction of 1 foot so the total adjustment for the entire vertical is only 1 foot. That is not at all out of the realm of possibility due to the movement of the boulders or the placement of the measuring device!

Something is happening with these calculations! Looking at the horizontal line measurements of each side of the cross as having a distance of 360 feet, when added together produce a sum of 720 feet, an even number. Again, using the number 144 as a divisor into the number 720, we get the whole number quotient of 5! The cross discovered by Fred Nolan has a mathematical design utilizing the numbers 3 and 5, two of the Fibonacci numbers used to make Phi! When the cross is adjusted for movement, either due to settling or the mere placement of measuring devices on the boulders, it is evident that the cross was not a natural occurrence but was designed by an extraordinary architect! This design was the work of someone who was a genius for their time! Who could that be is yet another mystery needing to be solved! But the evidence of the Fibonacci sequence and the Golden Ratio numbers being used in the design of the cross does give a clue as to when the cross was not made! Since Fibonacci wasn't born until the year of 1170 and did not publish his theory until the year of 1202, the cross could not have been designed using his concept any earlier than that date!

The next mathematics exercise was to use all of the numbers added together to test that outcome. Starting at the top of the cross, adding 145, 429, 293 produces a sum of 867, an odd number. If that number is divided by the first odd number of 3, the answer is another odd number of 289, an answer with no remainder. Now, take the number 360+360 to get 720 for the horizontal cross beam to get an even number answer divided by the first even number divisor of 2 to get what we added 360. Nothing genius about that one, but let's try the number 3 divided into 720 and we get 240; how about trying the number 5? The number 240 divided by 5 equals 48, an outcome of no relevance. Can we try the number 3 divided into 48? Yes and get the number 16, another even number all divisible by the number 2. How about adding everything using 720 for the horizontal cross and 867 for the vertical line to get a sum of 1587, an odd number total. If 1587 is divided by the first odd number of 3 the answer is 529, an odd number with no remainder, but try to divide that number by 144 and the outcome is 11.02, a non terminating odd number. Fibonacci math does not work on that outcome! If the measurements though are all changed to the Fibonacci number of 144 then an outcome by either multiplication or dividing the measurements of the cross can result in a number that has an even number outcome! More about this is next.

After analyzing the numbers of the cross, it was time to apply my Fibonacci hypothesis in believing it will unravel the code needed to decipher the cross! Looking at the cross, change the top measurement of 145 to 144, then change the other verticals below the cross beam from 429 to 432, and change 293 to 288 ft. These changes are made on the calculation that they are all multiples of the Fibonacci number of 144! All of these corrections now add up to 864 an even number divisible by the first odd number of 3 to equal an outcome of 288, an even terminating number divisible by 2 that leaves a quotient of 144! The number 2 is a very important number that I will explain later as it has a strong connection to the mystery of the cross.

In trying not to bore the reader here with any more arithmetic juggling, I will just cut to the point: there are two divisors or multipliers that keep popping up in the arithmetic, the number 3 and the number 5! So what do those two numbers mean, if anything? Let me explain.

What this indicates again, or actually proves, is that the Fibonacci sequence was incorporated into the measurements used in the design of Nolan's Cross with the hidden purpose of concealing some kind of a cryptic mathematical code! As explained above, in the Fibonacci sequence if the first number is divided into the second number in the sequence, the outcome is Phi or 1.6... – the Golden Ratio. So taking the two numbers found above as the divisors used, and divide the number 3 into the number 5, the answer is 1.6666. The Golden Ratio is produced which is also known as the Divine Proportion!

So how are these two concepts used to solve what the cross is hiding? There had to be something in the design of Nolan's Cross that all of this math was referring to that was just not evident at first or everybody would have discovered the same outcomes as I had. It was necessary to focus on what was already found in these outcomes of playing around with the math used for the dimensions of the cross. Once again, it required going back to the first step in recognizing that the cross was not proportioned in the normal layout of a Latin styled cross! The cross also had an extra boulder

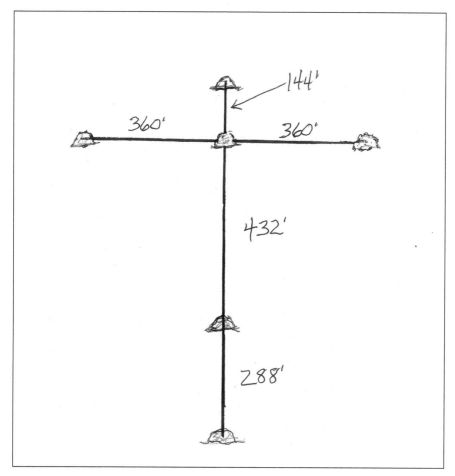

Nolan's Cross with adjusted measurements
Illustration by the author

placed in the vertical layout that is not usually needed to distinguish a cross image. All of theses things had to be considered together at one time to understand any connections, it did take awhile to realize that it was time to think outside of the box, and that was when there came a need to look even more closely at the cross!

What if there really was another shape disguised as a cross? There had already been attempts by others to use the cross as a base design to create something it was not. Peter Amundsen offered the theory that the cross represented the Tree of Life, also known as the Kabbalah. What Amundsen then showed was the geometric design of the Kabbalah overlaid on the cross that contained the 10 different points of Jewish belief, which is complex and not necessary to explain here. Another part of his written theory also involved a connection to the cross by Shakespeare! I give Mr. Amundsen credit for being so creative and detailed in his basis for the theory that was quite interesting. However, his is not a conclusion that I can accept. Mine is totally different with factual evidence that is forthcoming!

There were other attempts by different theorists that used boulders in the cross, located nearest to the swamp, to create a star shape with one of its end tips reaching into the swamp, claiming that it was pointing to the spot of the treasure. However, upon investigating their theory, there was no treasure found. I am certain there have been others who have presented their differing theories, but have probably been disregarded as just being too "off track" to even mention on the History Channel TV show!

What could the cross be other than a cross?

Earlier in this book, the chapter about symbolic clues described a few geometric sacred designs that have been used by civilizations throughout history.

Some of those early designs included basic shapes like triangles, circles, and polygons of all types that were combined to create an almost infinite array of special shapes with many being worshiped as sacred. It would not be necessary to begin to describe all of them as only a few are needing to be referenced here as being the most important. There are plenty of books available to the public that contain many geometric patterns with excellent descriptions of shapes that most people have never known to exist or have ever seen.

One of the first shapes to have been used by early humankind was the circle, representing the image of the moon as found on many cave walls. The circle is also believed to represent the Creator as being all encompassing, with influence over all that exists. There are other meanings about what a circle can represent but is often considered as being sacred and infinite as is the universe. As the ages passed, this shape took on many other meanings as it was manipulated or combined to create new shapes. Creations of the circle would be overlapped by other circles or would be manipulated into arcs to form what is known as a triquetra, a combination of three intersecting arcs pointing in three different directions like a triangle. The triquetra was believed to represent the Holy Trinity: the Father, Son, and Holy Spirit. This shape can also be seen in plenty of early architectural designs beginning around the 7th century. Another sacred combination of circles and the triquetra is the shape known as the Seed of Life or Genesis Pattern, both meanings are of the seven days in the beginning. This shape is a circle with seven interior circles linked together by a center circle with six arcs centered inside the center circle. This shape is often seen as a combination of circles found in many Christian church window designs in the shape of the cinquefoil, usually seen as a five-sided interlocking grouping of partial circles that represents the Rose of Venus. It is evident that the circle has played an important part in human history, and these are just a very few of the many scared examples that exist! Does a circle come into importance as to the cross found on Oak Island? Yes!

The next shapes to look at are the triangle and star. A triangle is any three-sided closed flat shape with its angles adding up to a total of 180 degrees. The triangle is one of the very basic polygonal shapes used in geometry. There are three classes of triangles as defined by their interior angles: obtuse triangles that have one angle equaling more than 90 degrees; acute triangles that have all three angles less than 90 degrees; and right triangles that have one angle exactly at 90 degrees. Other triangles can be classified by the length of their sides as being either scalene, equilateral, or isosceles. Some of the very first visual images recognized as triangles by most people are the pyramids found in Egyptian history. They were built with the sole purpose of being a religious burial site! Earlier, there was the description about ratios in math with one about the concept of Phi, the Golden or Divine ratio that was actually used in the construction of the pyramids. How it was applied is seen in the type of triangle that the pyramids are designed as and that is like an isosceles that has two equal sides. The formula used in the height of the pyramid is about the square root of phi or the ratio of the height is half the length of the squared base. Another use of the triangle can be seen today on the U.S. one dollar bill as a pyramid with the all-seeing eye of God on its top as it watches over humanity and is referred to as the Eye of Providence.

When King Solomon ruled in Jerusalem before the era of Jesus, Solomon used a ring as a signet with a design that used a circle and two triangles overlapping each other in opposite directions that created a hexagram inside of the circle. That ring was used by Solomon as his seal or stamp indicating him as ruler. There is much written by many authors and Jewish historians as to describing this item, which is usually referred to as either the Seal of Solomon or the Shield of Solomon. In Arabic mythology, the ring was supposedly given to Solomon by God with it having magical powers to rule over good and evil.

The equilateral triangle is often used in many sacred geometry shapes and is one of the shapes that was used early in history by Christians to represent the three parts of the Holy Trinity. But even before the Christians used the triangle to represent the Holy Trinity, the triangle was used in India as a sacred symbol representing the Shatkona. The shape of the Shatkona is made by overlapping two equilateral triangles facing in opposite directions to create a hexagram! In the Hindu religion, the two triangles are called the Shiva and the Shakti. The upward pointing triangle called the Shiva as being masculine and the downward triangle known as the Shakti being feminine with both triangles representing the divine union of male and female in all creation by God.

When looking at the Seal of Solomon and the Shatkona, both have the shape of a hexagram that is also seen as a six-sided star. The hexagram shape has been found to have been used as early as the Bronze Age of 3000 BC to 1200 BC The star shape has been used in many ancient cultures as either being a six pointed shape, or as a five-pointed pentagram shape, and even as a nine-pointed star known as a nonagram that represents the Baha'i Faith. The term "star shape" has been used by many cultures since the beginning of human history where it has been mostly seen as a religious symbol. For Christians, the star has been viewed by many as representing the five wounds of Jesus and was a sacred symbol used by early Jews. That use by Jews could also be attributed to the Star of David symbol also referred to as the Shield of David or the Magen David. In the biblical times of King David, father of King Solomon, the two shapes of the hexagram and the pentagram shape of the star were synonymous and interchangeable with each other in their use by the people. In the earlier chapter about symbols, the Shield of David was described as possibly using the hexagram in its design as was described in an earlier chapter on symbols. How these two polygons have been used in biblical history cements their usage as both refer to having a religious significance. The Star of Bethlehem has also been seen as a divine symbol announcing the birth place of Jesus.

What these shapes share with the mystery of Oak Island is that there is a relationship between the cross, a sacred religious symbol, and all of these other sacred symbols in telling the importance of religious symbols for humankind. Nolan's Cross may be just as important in religion as are any of these other symbols!

What does all of this math and the description of the sacred geometric symbols have to do in providing a solution to the mystery of Oak Island? Plenty!

The solution is found in the special math used to construct the cross; it will redesign Nolan's Cross into a sacred geometric symbol! This is quite a statement, but it is actually true and was only discovered by thinking outside of the proverbial box! The sacred symbol hidden in the cross will unveil a connection to a journey that has spanned centuries in carrying a sacred secret!

To begin the solution to the mystery, it is necessary to look again at the math used in the construction of the cross. No need to go over every detail – only what is important. It is necessary though to remember that the Fibonacci Sequence number of 144 is the clue that leads to the solution. Next are the division numbers used to get the whole number answers to the number sets that were created. Those numbers are the 3 and 5 that were used as the divisors. Since the numbers 3 and 5 are also numbers that can be divided into each other as was done in the Fibonacci exercise to get Phi that equals 1.6... or the Golden Ratio. It is important to analyze the cross to see if there is a Golden Ratio cloaked within. At first look, there is no evidence of any ratio occurring in the shape itself, but it is only found after decoding the math used in the construction of the cross! In order to find the Golden Ratio suggested by the math, the cross needs to be visualized as being a shape that has something with that ratio. Here is the key: the Golden Ratio used in Nolan's Cross is also known as the Golden Rectangle! This term comes from the Golden Ratio being applied to a rectangle where the ratio of its length is to its width. A simple example is the common, three-inch by five-inch file card everyone has used. That ratio of 5 divided by 3 equals Phi or 1.6...!

The Golden Rectangle is often used by artist in their paintings; Leonardo da Vinci and Salvador Dali used it in their paintings of the Last Supper. In those paintings, which are seen as rectangles, the placement of the disciples and other objects are visually balanced to be pleasing to the viewer by using the Golden Rectangle concept. An important note here is that Leonardo da Vinci, along with his mathematics friend, the great Luca Pacioli, collaborated to publish the book entitled *Divine proportione* in 1490! That book combined da Vinci's artistic talent with his knowledge of geometry and Pacioli's mathematical genius using the Golden Ratio to define the Divine in art. Much is written about this, but it is not necessary to go into any depth about it here. What is important to know is that these two great Renaissance individuals may know more about Nolan's Cross than has been previously thought! That's just my thinking and another theory for the Oak Island mystery solutionists!

Now we begin to suspect that the solution to the cross was hidden in a very clever design of mathematics using the Fibonacci concept of numbers that when decoded would yield the needed clue to find the Golden Rectangle!

In looking at the cross, the viewer must visualize that there is a rectangular shape hidden in the structure of the cross. That rectangle must use the important divisor or multiplier key numbers of both 3 and 5 to create the perfect rectangle. When looking at the cross there are only two straight lines to attempt to do such a rectangle, either using the vertical line or the horizontal line. Remembering to use the corrected measurements for either line when dividing with the number 3 or 5, or multiplying the Fibonacci number of 144 by 3 or 5 to form a rectangle. It was evident to see that the corrected vertical line adding up to 720 could be divided by 144 to get the divisor number of 5, or 144 multiplied by 5 would yield 720. That shape, if put on paper using only the one horizontal line as a base, would not create a spatially correct Golden Ratio rectangle! But, in using the horizontal line at the top of the cross as a base line, and drop another imaginary congruent (same) line down exactly 432 feet to the place of the extra boulder, you begin to see a rectangle that has sides of 144x3=432 and a width of 144x5=720, a perfect Golden Rectangle! Should you decide to create the hidden symbol found in the cross then the scale to use is not feet because there is not enough space to do so unless you're on a large field. Use the downsizing conversion from feet to the metric measurement of millimeters that will resize it to fit a standard sheet of paper that is 8.5 inches by 11 inches and use use a metric ruler and a drawing compass to complete the design. If you are not proficient at geometric drawings, then don't be shy in asking someone who can assist in doing the symbol. Precision is of the utmost importance in recreating the sacred symbol or there will not be an exact replication. The exactness of the diagonal measurements are necessary to create the correct angles and distances from point to point, an absolute requirement. Remember, this is a very unique symbol that must be done with perfection.

The next approach was to visualize what geometric shape the cross could be designed as to create what might have the same image of any other known sacred symbol. At first, I imagined the hidden shape or symbol that could be the ancient sign of Solomon, the hexagram! After trying many drawings of the hexagram there seemed to be too many unconnected lines that either needed erasing or would just complicate the drawing with inaccuracies of geometric exactness. However, there was something that became interesting when manipulating those attempts.

The cross was now beginning to create a very special shape that was similar to another discovered while doing research into sacred symbols. That other sacred symbol had a very deep religious connection to a hidden secret that could have an extreme impact on history. With the shape of the cross beginning to morph into this other sacred symbol, a realization began to overwhelm my thinking. What had just been discovered was bigger than anything I ever imagined! This discovery was so amazing that I could hardly believe it was actually hidden in the design of the cross, and in itself a spectacular creation done by some unknown genius from long ago! Is this really true that hidden in this arrangement of boulders forming a cross found on Oak Island actually is a unique symbol of only one known to exist in the world? Coudl this discovery alter what is believed about certain historical events? Could it be that the importance of this symbol is much bigger than anything that could ever be found on Oak Island? How would such a discovery be handled? The process of finishing the transformation of the shape needed to be completed to be certain about what was taking shape before my eyes! It was becoming an unusually unique perfect geometric shape!

Before revealing this very unique geometric sacred symbol, a little more background to its importance and relevance must be given. Be patient, please; it will be worth it!

The earliest of civilizations to use sacred symbols were known as the people of Mesopotamia who lived around 12000 BC in what is now Kuwait and Iraq. That region has been called the birth place of civilization. It was the people of this era who invented the wheel. One of the civilizations also emerging later from this era were the Babylonians around 2300 BC, who were some of the first people known to use sacred symbols in representing their religious beliefs and gods. The shape of a star was one of those first symbols used by humans that may have been a reference to the night sky. In Egyptian mythology, they used the star as a symbol to identify a sacred place they called the Duat which was where the soul would enter the afterlife to await resurrection. The Duat was an encircled five-pointed star that looked much like a star fish. There has already been mention of the star symbol used in Jerusalem in various ways, and a church currently found in Bethlehem displays one over the arched entrance that visitors walk under. There is also some research that notes in Jerusalem, around 200 to 300 BC, the use of the star symbol may have been used as an official stamp, much like Solomon did with his seal. History indicates the people of Jerusalem began using the star as a symbol of faith. The great mathematician Pythagoras who lived 570 BC to 490 BC believed everything could be described by mathematics stating the star/pentagram as the geometric symbol of perfection. The Pythagoreans used the five points of the star as representing the human body with two arms, two legs, and one head. Christians related the five points of the star to the five wounds of Christ. There are other interpretations used by some to suggest the five points as being earth, fire, water, air, and sky. No need to reiterate all of the meanings attributed to the star. It is a fact that the star has been used since the beginning of human civilizations to describe its importance as a cherished symbol representing sacredness.

There will be one more step needed to complete this shape before the final version is shown.

But before that happens, it is necessary to give a little more about the importance of the special number math used in the cross that tells the story about the Knights Templar connection. When people begin to think about numbers and what they mean, there is a reaction in the thinking process that identifies familiar numbers to certain things in life. For example, what do most people think when they hear the number 7? Probably the number of days in a week. How about 40? Possibly the number of hours in a normal work week or perhaps when you get to the age of 40 maybe it means being of a mature age! If 13, 16, or 19 are heard, we may think of the ages of teenagers. Numbers mean plenty in our daily lives, but how about when they are used as messages like the digits in a telephone number? Immediately, there is an identification as to who or whom that group of numbers refers to. Many people memorize many other number combinations or sets as part of their daily routine. How about locker combinations or street addresses, birthdays, passcodes, or bank accounts. Numbers are important because they convey something that is a useful message!

The measurement numbers found in the construction of Nolan's Cross are so important in not only telling how to redesign the cross, but they actually tell of another importance about who may have been behind that design. Please be patient in my explanation of this theory as it reveals through mathematics what I believe are the connections to the Knights Templar!

When the numbers were adjusted earlier to reveal the Fibonacci number of 144 as being critical to the development of the structure of the cross, there was one more number used in the arithmetic, the number 2. That was the intention of the designer as they had also hidden a message in the design that had to be decrypted. That decryption could tell who did the cross! All of this mathematical magic was definitely done on purpose as I believe the cross was designed to only be decoded when history would warrant its discovery. So everything found in the message was cleverly hidden by a genius of times long ago with intentions for its longevity.

Here it goes so follow closely as to what happens with all of these numbers as they begin to explain a hidden message cloaked in the mathematics.

The first number to decode is the 360 on either side of the cross, when those two sides are added together creating one complete line they equal 720. Using the number 2 as a multiplier, 720 times 2 equals 1440. Keep this number in mind for a moment. Next, take the corrected number in the vertical line that is 432 multiplied by 2 to equal 864, then take the other corrected vertical of 288 multiplied by 2 to get 576. Take both of those vertical numbers of 432 and 576 and add them together to also get 1440! There are two lines in the cross that equal the same when the math is adjusted to what is a code hidden by the designer of the mathematics! Back to the top vertical corrected to 144 multiplied by 2 to get 288. Add the 288 top vertical to the two bottom verticals total of 1440 to get 1728. Stay with me here. Divide the 1728 by 144, the Fibonacci sequence number and you get the number 12! Any clue yet? That's the number of disciples that accompanied Jesus! We are not quite done yet. Take the number of 144 of the Fibonacci sequence and when the digits in the number are added together of 1+4+4, you get the number 9, the number of the original Knights Templar that started in Jerusalem! How about this one: take the number of 1440 and add those digits together also to get 1+4+4+0=9 ! Take the total number of the corrected verticals of 1728 and add those digits together of 1+7+2+8=18 and then 1+8=9! Let's now try the 720 of the cross or even the multiplication of that number by 2 that equals 1440, and we still get a final sum of the digits as 9! Do you think there may be a hint in the mathematics used that may tell that the cross is related to the Knights Templar or even the disciples of the Bible? I do!

Here is another irony of the cross. In the book written by William F. Mann titled *The Knights Templar in the New World*, Mann displays a map of Nova Scotia with a drawn line indicating the distance of 1440 chains as the distance from Oak Island to a place known as New Ross to the north. Chains was the common method of linear measurement for long distances used in the 14th century. What Mann is describing in his theory is that this was the distance traveled by Prince Henry Sinclair, a Knights Templar who traveled to Nova Scotia in mid-1398. Notice that the number 1440 has the Fibonacci number of 144, perhaps a slight coincidence. Sinclair belonged to a family with ancestral ties to the Knights templar dating as far back as the First Crusade to the Holy Land! It was his ancestor named Henri de Saint-Clair who actually served with Geoffrey de Bouillon in retaking the city of Jerusalem from the Muslims! Written earlier in another chapter, it was Godfrey, along with a small group of other knights, who probably found the scrolls that were used by Ralph de Sudeley for his journey to Nova Scotia. Here are a few of the many pieces needed to fully understand the connections that have occurred through the centuries that must be fit together in order to see the complete picture. This is just one more of those pieces talked about in the beginning of the book as to why each chapter adds another piece to the solution of the mystery. I hope you didn't miss any!

There are clues in the math that are used to construct the cross that not only tell what the shape is supposed to actually represent, but the corrected numbers used to decode the math possibly also tell who designed the cross! Here is the connection in the math that could tell who might be the designer. The measurement of the horizontal cross being a total of 720 feet across reminded me about a historical event that I had researched involving a famous person in medieval history of 1502 that designed a bridge that was 720 feet long that was never built.

The bridge designed by this noted engineer of his time was commissioned for the purpose of crossing a river named the Golden Horn! Imagine the irony in that river's name and the ratio of the Golden Rectangle being found in Nolan's Cross that also spans 720 feet! This same person was also very knowledgeable in hydrodynamics even to the degree of designing the engineering plans for the diversion of water flow from a major river known as the Arno located just northeast of Milan, Italy. There was even a plan for a dam that would limit the flow of water into an area that needed a controlled water level. Sound a little familiar? Like maybe some of the necessary knowledge needed to create the flood tunnels on Oak Island? This person was also thought to be involved in some capacity with the Knights Templar! There is even a strong possibility that this person also had knowledge about the Ralph de Sudeley journey to Nova Scotia in 1178 as he had close acquaintances who had first hand knowledge about the *Templar Document* written herein.

Keep reading; the answer will be told!

So there just seems to be an ever growing number of participants functioning as co-conspirators who over the centuries have discreetly contributed in one way or another to protect some secret knowledge about what has taken place on Oak Island. There is no definitive record as to who has been involved as of yet, but maybe the math and geometry here will help. Maybe you the reader may have an idea?

It is time to complete the geometric symbol concealed in the cross. Should you attempt to do the design, be certain to use all of the corrected measurements that were given earlier and be precise! Begin by looking at the cross as having an imaginary rectangle in the middle of the symbol below the top horizontal cross beam of the vertical. If you have not already done so and are attempting to draw the hidden symbol, remember to drop an imaginary horizontal line down to the plane of the boulder below the cross headstone beam. You can, if it helps, lightly draw the side lines to connect the rectangle to fully recognize this shape. They will be erased later.

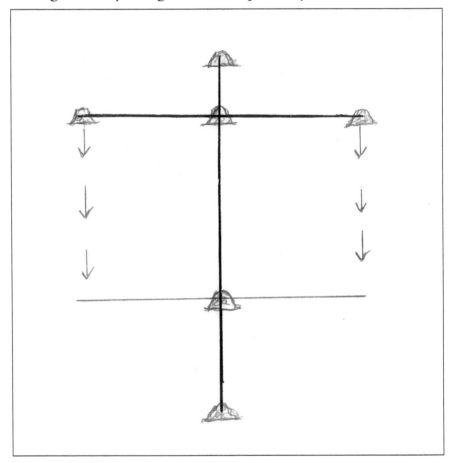

Forming the rectangle of Nolan's Cross
Illustration by the author

Remember, this must be done with precision or the shape you attempt will not be replicated! Check the diagonal distances from each of the opposite ends of the horizontal lines as they must be the exact same length or their angles will not be correct ! If the lines are not equal you may need to correct the angles of the horizontal lines to ensure that they are at 90 degrees.

If you want to construct the sacred symbol, then you can copy my cross without the arrows as it has the correct measurements to follow in the next steps.

First, draw a straight line from the bottom right of the lower

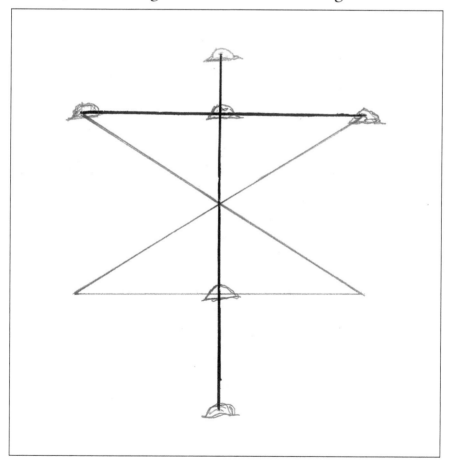

Nolan's Cross with the X
Illustration by the author

horizontal line end to the top left upper horizontal line end. Then draw a line from the bottom lower left horizontal line end to the top of the upper right horizontal line end. What this has done is the crisscrossing of the rectangular box that has now created an X in the middle of the shape.

Now, draw a line from the top right horizontal line end to the very tip end of the center vertical line at the bottom of the shape. Then draw a line from the top left horizontal line end of the rectangle to the very bottom tip of the center vertical line meeting the other lines at the bottom. Erase the center vertical line completely and erase any side lines of the rectangular box. Rotate the shape so that the bottom is now the top and the top horizontal line is now the base. You have found the STAR – a very unique sacred geometric symbol hidden in Nolan's Cross! Unbelievable!

When studying the shape that has been created, it is evident that it is a pentagram or star symbol, but not quite a correct mathematically described pentagram! A mathematically correct pentagram has five interior angles with each equaling 36 degrees that creates a star shape that when rotated looks the same from any angle! This Nolan's Cross star does not do that no matter which way it is viewed. It looks different at every turning. Plus, the interior angles of the star arms are not all an equal distance apart! If the distance between each star tip is measured on a regular pentagram they are, by definition, all an equal distance apart, but this is not so on Nolan's Cross. If you want to create a pentagram to see for yourself the irregularity of Nolan's Cross, just draw a circle using a compass, then mark off five points each at 72 degrees apart. Next begin to connect the points on opposite sides of the circle at every other point until you have a five-pointed star that is a regular pentagram. With all of the nonconformity to the correct geometric definition of a pentagram, it is amazing that a perfect Golden Rectangle is at the center of the symbol!

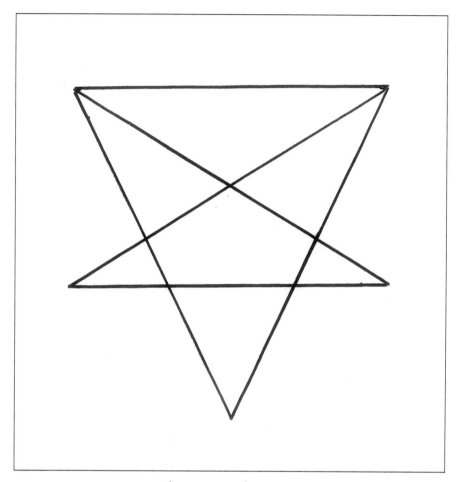

The Star in Nolan's Cross
Illustration by the author

It is now necessary to use a drawing compass to encircle the star to complete the hidden symbol! Open the compass legs by placing the pointed tip of the compass at the exact point where the lines crisscross in the center of the symbol. Next open the drawing leg of the compass so that it touches the ends of each horizontal line and secure this opening of the compass. This distance must be exact and the setting must be maintained. Using the drawing leg of the compass, begin to completely encircle the symbol touching each end of the horizontal lines as the drawing line passes over them. If this is done correctly, then a perfect circle is now surrounding the symbol as it has contacted each end of the two horizontal lines. Do not change the compass setting! It is best to double check the gap setting of the compass to be able to do the next circle correctly. Using the point end of the compass, place it at the very end of the bottom right horizontal leg and tilting the drawing end of the compass up to the right and touching the circle, lightly draw a line to the left making a small arc inside of the symbol toward the center X. Next put the pointed end of the compass at the top of the symbol point and move the drawing end of the compass inward to the left coming in from the right outside of the circle making a lightly drawn arc line toward the left that will crisscross the first arc. The two arcs must crisscross exactly as this is where the next circle will begin. It would be a good idea at this point to double check the opening of the compass to ensure the original setting has not changed. Place the pointed end of the compass at the exact location where the two arcs have crisscrossed to the right of the centered X. Tilt the drawing end of the compass toward the right side of the symbol and begin drawing a circle that will overlap the first circle touching the ends of the bottom right horizontal line and the top of the star as it is drawn around the symbol. Erase any crisscross arcs drawn inside of the symbol.

This is a very interesting symbol! Some may think that this symbol is associated with Wiccan practices or that of Satanism and the occult, but it is quite the opposite. This shape was designed long before those images were ever created! Remember, everything involving Nolan's Cross on Oak Island was done before 1795. Any previous usage with the pentagram was nothing other than being one of divine or holy intentions! The pentagram used as something other than a sacred religious symbol has only come about after 1897. Please dwell on that thought for a moment. Any use of the pentagram other then its use as a sacred symbol did not come about until after 1897 when it was given a different meaning after being

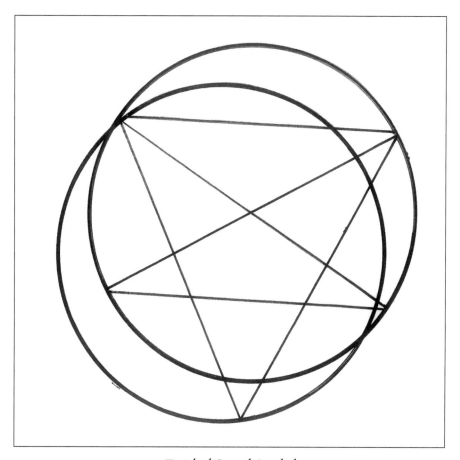

Finished Sacred Symbol
Illustration by the author

displayed in a book by Stanislas de Guaita, a French author and occultist! The pentagram he described was encircled by two unconnected circles with what appears to be a goat's head in the center area of a perfect pentagram. That shape has been associated with the Wicca, a Paganism modern religion that practices witchcraft and worship of different gods or goddesses which became popular as a cult following in the 1950s. That practice began in England and later used a version of the double encircled pentagram as a symbol that was promoted by the entertainment medias of movie and literature. The double, unconnected encircled standard pentagram is not the only symbol used by the Wicca. In the book *La Clef de la Magie Noire*, which basically means "The Key to the Black Magic," describes several different uses of the pentagram even including one similar to the Vitruvian Man of the Da Vinci art concept of proportional balance of subjects. Many symbols in history have similarities to others due to the great number of actual symbols in use. Even in World War II, the use by the Germans of the swastika actually had a totally different meaning than what they emblematized it to be. The Germans used the modern right facing swastika to instill fear and dominance over those they opposed in the war. The ancient left facing swastika was originally a religious symbol of divinity used in the Eurasia cultures of Europe and Asia. *The Vedas*, a religious text, used by the early people of India circa 1500 BC used the swastika symbol as one of divinity with spiritual meaning. Depending on the direction the swastika is facing changes its meaning. In the various religions found in other lands around India, the left-facing swastika is also revered as the symbol of divinity and goodness. In many other regions of the world, it is even viewed as the form of a cross and is considered by many as a symbol of good luck!

Unfortunately, in today's society, there always seems to be those who will twist anything good into something different just to fit their focused narrative and view without regard to the true view of others. Take a look at how some media platforms have twisted the facts in situations to make them more news worthy; they have even accused each other of being "fake news." Look at this very unique encrypted symbol that has also been found an ocean away at a location where it also has a religious significance. That's a little hint to the rest of the story in solving this mystery, but it is getting close to the time to tell. Accept this symbol for what it is: as a one-of-a-kind creation designed hundreds of years ago by a genius who used a mathematical code and geometric symbolism to encrypt a unique sacred symbol with a special message!

So how does society see the star shape or pentagram in our world today as a shape in history? Some pentagram importance in history has already been given earlier especially as to how it has been viewed by the various religions around the globe. Here is another important example why the pentagram has had such an influence in everyday peoples lives: the architectural display used in churches. Have you ever stopped to think why so many churches display the pentagram symbol in their art creations? It is because of the divine perfection associated with that symbol as identifying the structure of the church and its teachings as being divine! The pentagram symbol has the meaning of divinity!

Throughout history, the pentagram has been found displayed in many churches not only located in the Holy Lands, but in other countries like Portugal, Spain, Italy, Germany, England, and others.

A modern day example of the pentagram symbol used by those of goodness can be found on the Church of Jesus Christ of Latter-day Saints Mormon Temples! This religious association was founded in what could be considered modern day time of 1830 in Fayette, New York. Their churches in Salt Lake and Logan, Utah and Nauvoo, Illinois use the pentagram symbol on their places of worship in taking a reference from the Bible in Revelation chapter 12, "And there appeared a great wonder in heaven; a woman clothed with the sun, and the moon under her feet, and upon her head a crown of twelve stars."

Let's take a brief look around the world at how pentagrams have been used in some other places that have used divine architecture before the reveal of what Nolan's Cross is messaging.

The Church of Santa Maria do Olival found in Tomar, Portugal has a pentagram boldly displayed within a circle on one side of hte church. Here again on the side of a historic church is found two sacred symbols indicating a structure to be seen as a divine place. That church happens to be one of the churches built by the Knights of the Order of Christ, otherwise previously known as the Knights

Templar. The 12th century Templar Master of Portugal, Gualdim Pais is buried inside this church which also displays a statue of the Virgin Mary and Child. Surely this location is recognized as being sacred!

Next is a church found in Lisbon, Portugal known as the Lisbon Cathedral or Roman Catholic Cathedral of Saint Mary Major. The church was built in 1147 after the city ousted the Arab controllers by soldiers of the Crusades. There, displayed on the outside of the cathedral, is a pentagram with intertwined lines shown inside a circle. The circle is believed to be sacred by several religions as in the meaning of the "Circle of Life" – there is no beginning nor end. With the pentagram being encircled, again it represents another double sacred symbol as the circle was the first symbol in ancient times to be regarded as sacred. The circle represented many things: first, the moon and the sun, and later, the all-encompassing love of God that surrounds everything.

Another church found in Portugal with the pentagram displayed in the archway of its entrance is the church in Loule, located at the southern tip of the country. There are many churches in this area, but some have only their remains left after a severe earthquake occurred in 1755. Portugal had many more churches early on in its history, more than what one would expect for the size of its population. Maybe it had something to do with the Portugal name that actually means "Port of the Grail."

The Market (Marktkirche) Church, in Hanover, Germany was built in the 14th century has a large pentagram within a circle on one side of its structure. A pentagram within a circle is called a pentacle.

St. Barnabas at Bethnal Green in London, England, built in 1865, displays a very large, perfectly styled pentagram on an outside wall.

The most known church to all is the Notre-Dame de Paris. That name actually translates into "Our Lady of Paris" due to its dedication to the Virgin Mary. The building of the cathedral began in

1163 and was erected using the engineering method of rib, vault, and flying buttress assembly that created a very impressive architectural masterpiece. That type of construction allowed the cathedral to be built at a much taller height than those that had been done previously. The church has several rose windows which are designs of rose petals that are very colorful. But what is most impressive at this cathedral – for this book topic – is the addition of the North Transept in 1250-1260. A transept is a part of the church built like a large arch enclosure usually found at the ends or sides of a church.

Amiens Rose Nord
Public Domain Image

The transept found at Notre Dame was built showing a large window designed with a pentagram that tilts slightly down on its left side, very much like the cryptic symbol of Nolan's Cross! This particular rose window in the North Transept is known as the Amiens Rose, a very intricate and beautiful masterpiece of geometric art! In the view of those familiar with sacred symbolism, a left-leaning pentagram supposedly represents the eyes of God looking down upon the Earth. Try to remember this one as it will come up later. The cathedral also was the location of relics thought to be from the passion of Christ such as the crown of thorns and a piece from the actual crucifixion cross!

Another very impressive display of sacred symbolism is found at the Templar Hermitage of St. Bartholomew of Ucero. This church was Templar-built in the 12th century in Soria, Spain. The church is located in an area known as Ucero Rio Lobos Canyon which is found in the northeast region of Spain where limestone caves are hidden among the low-lying mountains with steep cliffs. Templar activity was prevalent in this area during the 12th century serving as a location of refuge that once had been used as a spiritual site during the days of the Romans. A statue of St. Bartholomew can be found here, recognizing him for being one of the 12 apostles chosen by Jesus. It has even been suspected that the Hermitage was built using symbolism indicating the possible location of the Ark of the Covenant. The rose window on the hermitage is a beautiful design of the star or Pentagram that is slightly leaning to the left. Yet another display of a sacred symbol to those who understand its meaning!

A few more churches to mention where the pentagram is displayed are the Church of San Juan in Castrojeriz Burgos, Spain, St. Francis Porto, Portugal, the Cistercian Abbey in Hauterive, Switzerland, and St. Peters Cathedral, London.

The importance of the pentagram is its historic use as a sacred

The Templar Hermitage of St. Bartholomew of Ucero
Note the symbols in the window.

symbol that has been found on artifacts and buildings that date back to the times of the ancient civilizations of the Egyptians circa. 3100 BC, and even before that time by the Babylonians. The Babylonians used the pentagram symbol as representing their belief in a god they called Ishtar, the goddess of love. The Druids, a religious class in Celtic times during the 3rd century BC, worshiped many gods and goddesses and believed the pentagram represented their Godhead, their supreme deity. The symbol had been seen as representing something to venerate; it meant goodness and was often used to ward off evil when it was displayed on buildings and artifacts. At first, the symbol may also have been thought to acknowledge the five elements of life; earth, water, fire, air, and spirit. Christians, after the death of Jesus, have likened the pentagram to the five wounds of Christ: two wounds to his legs, two wounds to his arms, and one wound to his side. Another use of the star has been to reference the various directions of the earth: east, west, north, south, and above.

Pythagoras, the ancient Greek Philospher and renowned mathematician who lived 570 BC-490 BC, gave us the Pythagorian Theory, a basis to Euclidean geometry. The theory states that the sum of the squares for the legs of a right triangle are equal to the square of the hypotenuse, the long diagonal side of the triangle. The Pythagoreans also used the pentagram symbol to represent their identity and even equated the symbol to their goddess Hygeia. Pythagoras believed that the pentagram was the most perfect geometric symbol of all. The cross that exists on Oak Island morphs into a pentagram because of the Golden Ratio or the Divine Proportion, a perfection in mathematics that is found in its cryptic design!

The second part of this cryptic sacred symbol is the encircling of the pentagram that basically creates what is known as a pentacle. This pentagram is encircled not just once, but twice. A double overlapping circle symbol in geometric terms is known as the vesica piscis, a shape found in many religious settings as a divine configuration. But first, the circle must be given its importance as being a sacred symbol on its own that has been worshiped since the beginning of human time. One of the first beliefs about the circle is that it is a very basic sacred symbol as viewed by human kind as thought to be the symbol that represents the concept there is no beginning nor end, and that everything exists in some form for eternity and never terminates. The circle was also seen as the shape of the moon and the sun with both worshiped by early civilizations naming their gods and goddesses as having special attributes. Both the sun and the moon were permanent objects to be seen in the sky giving the ancients heavenly objects to worship above and were seen as being divine. It seems like every religion on earth had a named moon or sun goddess. Just pick almost any country or religion and you can find the name of a goddess that will range in the alphabet from the letters A to Z like Anahita or Zelena/Selene, the ancient goddesses of the moon! Anahita was a Persian goddess of the moon whose name meant "one of purity." Another very familiar moon Goddess is Luna; in ancient Rome, she was the image of the moon, and her name is the root for the word "lunar," a modern term used for the activities related to the moon. There are actually dozens of named goddesses found in mythology to be either Greek, Egyptian, Hindu, Buddha, or even by the Indigenous peoples of the various continents of the earth. The ancients observed that the moon had its changing phases or cycles that repeated monthly like the human female who also has a similarly timed monthly cycle. With the feminine cycle and the moon having similar time cycles, it is easy to understand why the ancients equated the moon as being a feminine object. The main point here is that the circle represents a worshiped shape as being divine or sacred. The circular shape of the moon was easy for the ancients to describe the moon as a goddess, a divine shape seen as a feminine deity.

We cannot forget the sun goddesses and gods. The sun, of course, is the other sacred circular shape seen in the sky that has also been worshiped since the beginning of time. The sun is seen as the energy that makes things grow on earth and that also provides the warmth for all to absorb. The sun is also the light that lets all things see their world. Again, the number of sun gods is as many as there are human cultures around the globe. Ever since humankind has looked up to the sky to see that big, bright circle, there has been the worship of its shape as being divine. A few of the more popular sun gods range from the great Egyptian mythical sun god Ra to the sun god Huitzilopochti of the South American indigenous Aztec Indians. There are others like Helios, who was worshiped by the Greeks, and Surya, worshipped by the Hindu. The native American Indians also had their sun gods like the Navajo, Mohave, Zuni, and Seneca. Sun gods were worshiped as being strong, masculine, bright, dominate, magnificent, and were the most important one of all to some cultures. In Christianity, our God is still there even when it is dark. He is larger than anything we can see but yet not touch. His light that shines give us warmth and shows the way. He was here in the beginning and will remain beyond our time. God is eternal, just like the circle with no end. The circle is and has been a symbol of sacredness!

Let's get back to the vesica piscis. The joining of the two circles has also been seen as the union of the male and the female, just like the wedding rings used in marriages that represent the joining of husband and wife. Like the circles, with one representing the feminine deities of the moon and the other representing the masculine deities of the sun, each revolve around the one most important symbol of the earth. In a marriage between two individuals, their union is to be sacred, revolving around the holy vow they took before God. The point is the importance of the circles being joined or overlapping in creating a sacred union in the geometric symbol of the vesica piscis. The shape has also been described as being the mother of geometry in that it not only is a circle, the most basic of shapes, but also contains triangles which are the parts of a square.

By now, you have hopefully gained an understanding about the importance that symbols have played in human history ever since the beginning of time. The art of geometry and the precision of mathematics has been used to create that art form in a way to present something that is both recognizable and meaningful on its own. Geometric symbols have been used as art forms in the beautiful depictions of worshiped gods and goddesses throughout the centuries such as in the Hindu religion with their very colorful and complex chakras. Flags around the world display the star symbol of the pentagram. It is often worn as a badge by law enforcement like the one worn by the U.S. Marshalls. The U. S. military Medal of Honor, one of the highest recognitions offered is shaped as a pentagram. The official seal of the United States is surrounded by many stars that are also boldly displayed on the U.S. flag. The importance of the pentagram/star has forever been a part of daily human experiences with its display as a reminder to its perfection. The star has been revered through the ages as a sacred symbol! A simple reminder about one of the most important stars known to early ancients was the one seen in the sky to signify the birth of Jesus Christ, the Star of Bethlehem. Every Christmas, people around the world decorate for the holiday that celebrates the birth of Christ. A lot of decorations involve the use of some type of tree that is often adorned with a star/pentagram on top. That adornment is seen by Christians as an image of the Star of Bethlehem above the gift below which was given to us! Sometimes we forget the powerful meaning of a sacred symbol.

Nolan's Cross has been concealing a sacred symbol within its design ever since the boulders were first positioned so precisely, clearly it was done with some very special purpose in mind! But why would such a thing ever be done? Could it be that in the expansive spacing of the giant boulders there was the intent to conceal the discovery of the cross and what it hides? But why? Could it be that the hidden sacred design found within Nolan's Cross was never meant to be known? What secret could be so valuable that a cryptic sacred symbol is its key? There must be something more to the meaning of this geometric symbol than just being encrypted there. Why is it that this particular Christian cross can be transformed into this very unusual one of a kind very sacred symbol? The only reason that makes sense is that there must be more of a story yet to be unfolded! Could it now be thought that the uncloaking of this hidden symbol could lead to the real mission of why Oak Island exists, and that this is only the beginning of something even bigger? It is true that this is only one piece of the puzzle.

Take note: this is not the only unique sacred geometry symbol like this. It has an exact twin in the homeland of the Knights Templar. These are the only two symbols known to exist today in the world! What do they mean?

It is time to finally reveal the reason this book was written. Not only was the decoding of the Oak Island sacred symbol an extreme discovery of great importance, but the knowledge of another symbol warranted their disclosure to the world! It wasn't until I had deciphered the encrypted pentagram in the cross that there was the realization that there are two identical sacred geometric symbols that have been cleverly disguised for history to solve their meaning and their secret!

This story now goes much deeper than one just about those suspects who were thought to be hiding a cache of gold and silver or even about the many who have sought to retrieve the possible treasure of immense wealth. The twin sacred symbols are now puzzle pieces to a more far-reaching mystery that could have a much greater reward than any imagined treasure of gold or silver!

The discovery of the sacred geometry symbol cloaked in the mathematics of Nolan's Cross has revealed a greater and more divine reason for the genesis of the Oak Island mystery. What could it be?

Are you ready for what comes next?

The Sacred Symbol Connection!

There is only one important treasure to be found on Oak Island, and it is Nolan's Cross! The search for any treasure was begun long ago by a trio of young men named Smith, Vaughn, and McGinnis, who discovered something suspicious on a small island off Nova Scotia in the year of 1795. But that was just the beginning of what has become a very complex mystery that continues to this present day with more questions than provable answers. That treasure search has now been pursued by many for more than 200 years without reward of any riches. The mystery continues to baffle those who have tried to unwind the secret that tells the story of who, what, when, and why. There is an answer to the mystery, though, but it is not found in the legends of pirates or the suspicions about some covert operation that was led by adventurers who went rogue in an effort to stealthily hide a cache away from either royalty or country. The true riches of this treasure will be found in the reward of unraveling a story so significant it could alter history with discovery of the encrypted sacred symbol cloaked within Nolan's Cross!

The discovery of that symbol was totally unsuspected until it was found by a willingness to search beyond the quest for a rich treasure of precious metals. The sacred symbol reveals something more precious than what any amount of gold or silver will ever enrich the recipient. It leads to the enlightenment about some intriguing events in a time that history has left as ambiguous, obscure, or even suppressed!

Several hundred years ago, something uniquely mysterious occurred on Oak Island that has a direct relationship to another intriguing event that occurred an ocean away near the small commune of Couiza, France. That is the location of a small church known as the Rennes-le-Château, a place with its own strange mystery. That church can be found in the French region of Languedoc located in the most southern area of France at the foothills of the Pyrenees mountains. This once meager church that was originally built in the 8th century, was rebuilt in 1059, and once more in 1891 by its then priest named Bérenger Saunière. His story has now become entwined into the mysteries of Oak Island. How could two things separated by an ocean with one event in France have a connection to Oak Island?

Several books have been written about Sauniere's activities as to how his knowledge about something he found at this little church actually helped to reward him with substantial wealth! The mystery stems from this: during the renovation of the church in 1891, some hidden parchments were discovered that had been written in old Latin text containing an encrypted message. These parchments supposedly had information of a treasure that could be of great wealth or of Biblical importance! Maybe even both! Whatever that text contained was so important that the Catholic Church wanted to be its possessor and wanted to keep it secret.

The discovery found at Rennes-le-Château included four different parchments. One of those parchments contained certain letters that formed an encrypted geometric sacred symbol that is the twin to the one I decoded in Nolan's Cross! Let me repeat that statement:

the encrypted sacred symbol in the Rennes parchment is the exact twin to the encrypted sacred symbol that I decoded in Nolan's Cross and there are no others like them anywhere!

The discovery of these two twin symbols could lead to the most significant historical and religious finds in our time.

How can the connection of these two encrypted symbols explain the mystery of Oak Island?

It is the back story that reveals the importance of their combined message. That message is one that has been developing for centuries about events that have involved many participants, but has always revolved around one objective: to hide something so important that it could change the history books!

First, there is a need to give some background about those events and to give credit to a couple of individuals who also are a part of this fantastic discovery. What makes Rennes-le-Château so important in this story is that it was the location of much activity in France during the era of a certain religious group in the 13th century who were known as the Cathar. This religious group was self-described as being Gnostics, those who claimed to have special knowledge of spiritual things. Others condemned them as being sinful heretics who were seen as reclusive with strange rites giving cause for the church to proclaim their condemnation. There were several communities of these believers who were not followers in the beliefs of the Catholic Church and had their own opinions about biblical teachings and its history. There was an effort by the then-Pope in 1244 to militarily eradicate any Cathar existence from the regions in southern France, especially at their commune located at the château on Montsegur. That mountain commune of the Cathar stood at about 2800 feet high and was located just over 30 miles away from Rennes-le-Château!

The Cathar, though, had been accepted by the Knights Templar who were well established in what was their home area of the

Languedoc region during that same period of time. The Knights had established several commanderies in the neighboring foothills of the Pyrenees near Montsegur at Le Bezu, Laval-Dieu, and possibly at Mt. Cardou along with others. The Knights Templar, who on an occasion or two, had been rumored to have even driven away enemies of the Cathar. According to legend, the Templar had a reason to be protective of the Cathar. It had long been a belief that not only were the Knights Templar protectors of the Holy Grail and the Ark of the Covenant, but that the Cathar either had been given possession of those items or actually held the knowledge to the secret location of those holy relics!

The Cathar would unfortunately suffer their fatal demise at Montsegur in the year of 1244. That is when the soldiers of the French king, with approval from the Pope, captured the commune of the so-accused heretics and condemned more than 200 of them to be burned alive in a large pile of brush, twigs, and branches known as a pyre. The fate of the Cathar had been that they would not accept changing their ways, defying the desires of the Catholic church to submit to Catholicism. The Cathar believed they were the only true believers in Jesus Christ and the obedient followers of his teachings, and as such, they would not accept the dogma of the Catholic Church. The Pope was also fearful that the Cathar would undermine the church's influence, possibly causing discord among the followers of the spreading Catholic theology. They had been labeled as an enemy of a powerful authority.

The irony of all this is that the Knights Templar would eventually also suffer similar accusations of heresy that were brought by the King of France and the Pope later in the 1300s. The Templar Order who had once been the heroic servants of the Catholic Church during the times of the Holy Crusades would now become pursued, imprisoned, tortured, and killed by the soldiers of the French Army, which was all approved by the Pope. The Pope, under duress from the King of France, would support false accusations of heresy against the Templar declaring them as enemies of the church. All of this was a false scheme contrived by the King, who owed a great financial debt to the Knights and needed an excuse to dissolve his financial commitment! The Knights accumulation of immense wealth and status had made them a target of two unscrupulous persons of power who together would sanction the Knights downfall!

There is a relationship in history that binds the Cathar and Knights Templar groups together that both were suspected to have knowledge or possession of the Holy Grail and the Ark of the Covenant. Before the last days of the Cathar atop their refuge at Monsegur, it has been written that a few of the men escaped the condemned commune and carried something of great value away to be hidden elsewhere. That elsewhere was thought to be Rennes-le-Château, a meager little church that had been dedicated to St. Mary Magdelene!

The southern region of France had been the cradle of the Knights Templar when it began back in the 12th century, emerging through the years of conflicts in the Holy Crusades to become the mightiest of medieval armies. However, the Order was dissolved when the last Grand Master Jacques de Molay was falsely accused of heresy by the King of France and the Pope who condemned him to be burned alive at the stake! The Knights Templar, however, were not eradicated from Europe as the King and Pope had hoped. Many Knights had evaded capture after being forewarned about their pending demise on Friday the 13th of October 1307, a day of infamy for France!

If there was something of great religious importance the Knights Templar were in possession of, would it not be reasonable to think that it could be hidden in that region where both groups knew as their homeland? What Bérenger Saunière found at his small church during the renovation in 1891 may have been the secret knowledge that had been protected for centuries by these two groups.

Whatever the discovery, it was important enough for Saunière to travel to Paris to present those once hidden parchments to experts of the church for them to decipher what had been written in old Latin text. It was told that one of the priests quit the church after reading what Sauniere had brought.

What could have been written in those parchments to cause such a reaction? The content of what was written in those parchments was not revealed; in fact, some of the parchments were never returned to Sauniere but were kept by the church. An interesting circumstance did follow the visit of Saunière to Paris; he suddenly acquired a great deal of wealth which he spent freely on his little church, decorating it with expensive things of personal taste and constructing a new residence. It was also said that he hosted many lavish parties for people of notoriety, something not the usual habit for a low level priest with a supposedly meager stipend. There is a lot more to his story that could be written here, but what may be important to ponder is that on his deathbed, one of his friends who was a priest would not offer him absolution after hearing Saunière's final words in his confession about what the parchments contained. Why?

This amazing story about Saunèire, the little church of Rennes-le-Château, and more can be found in a couple of books, including one written by the late Sir Henry Lincoln titled *Key to the Sacred Pattern* and another by Gerard de Sede titled *The Accursed Treasure of Rennes-le-Château* originally written in French. The late Sir Henry Lincoln is the person who discovered the hidden sacred symbol in 1971 that had been encrypted in the parchments found by Sauneire. I would love to display that page in his book with the parchment overlaid with the sacred symbol as was done by Sir Lincoln, but because of copyright laws, it was not possible to obtain it before his passing in early 2022 at the age of 92. However, his depiction of the sacred symbol, which is an exact twin of the Nolan's Cross encrypted symbol, can be found online displayed on the front cover of his book as well as inside. Sir Lincoln also authored many books including *The Holy Blood, The Holy Grail, Saunière and the Decoding of the Mystery of Rennes-le-Château, the Holy Place*. He was also a long time writer and producer for several TV series that aired on the BBC in London. Sir Henry Lincoln was a highly talented author, producer, respected person, and a very smart individual!

How all of this comes together starts with the initial discovery by Sir Lincoln in his deciphering of the sacred symbol in 1971 which was hidden in the parchments uncovered by Sauneire in 1891. With the discovery of the exact sacred symbol that I was able to decipher in the geometry of Nolan's Cross, it is beyond any doubt these two symbols tell a much bigger story that is connected to what happened in France centuries ago! The mystery of Oak Island now goes even much deeper into ancient history as it has become entangled into the mysterious web of Rennes-le-Château and beyond!

For those who have been wondering who was the genius mathematician that calculated the brilliant geometrical design of Nolan' Cross that incorporated the unique sacred symbol, it was in my thinking Leonardo da Vinci! If you missed it, go back to the last chapter and read about halfway through to see the paragraph with the information about someone wanting to build a bridge span that was 720 feet across in Italy. Leonardo didn't get to build that 720 foot design in Italy, but I bet he knew how to create a 720-foot cross! Leonardo was friends with a very smart mathematician named Luca Pacioli, the father of modern accounting. Leonardo collaborated with Pacioli to write a book about mathematics and illustrated the Divine Proportion/Golden Rectangle in that book. It's all about the math and geometry that has been used in this mystery. Leonardo was born in 1452 living until 1519; those are the years that fit very well into the time frame of artifacts found on Oak Island. Maybe Leonardo didn't come to the island, but someone used his genius to complete the mission.

```
        ℳ
        ⳨    ETFACTUMESTEUMIN
SAbbATOSECUNdEPRIMO A
bIREPERSCCETESdISCIPULIAUTEMILLIRISCOE
PERUNTVELLERESPICASETFRTCANTESMANTbUS + MANDU
CAbANTQUIdAMAUTEMdEFARISAEISdT
CEbANTELECCEQVIAFACIUNTdTSCIPVLITVISAb
bATIS + QUOdNONLICETRESPONdENSAVTEMINS
SETXTTAdEQSNVMQVAMbOC
LECISTISQUOdFECITdAVTdQVANdO
ESVRVTIPSEETQVICVMEOERAI + INTROIbITINdOMVM
dEIETPANESPROPOSITIONIS        REdIS
MANdVCAVITETdEdITETQVI        bIES
CVMERANTUXUS QUIbVSNO
NLICEbATMANdVCARESINON    SOLIS SACERdOTIbVS
                                      (P S)
```

Sauniere's Small Parchment
Image courtesy of Allysha Lavino

One more echo of importance: there is some form of treasure on the island! It has to do with what was once at Rennes-le-Château!

If you are still curious as to who put the cross on the island, there has always been the one group with the ability to execute something of religious importance who could also have possession of something needing the utmost secret protection. It has to be the Knights Templar or their successors! Right?

Or could it be another group of religious devotees? What if it was the church itself?

It sounds a little absurd, but maybe not really when you think about who identifies themselves with the cross? The church! The Knights Templar actually uses the "la croix pattée," a symbol similar to a cross, but not that of a Latin style cross as is Nolan's Cross! The church would also have enough influence to assemble any necessary talented group of highly devoted Christians to secretly do whatever the church would ask! Those groups could have been ship owners with devoted crews as laborers, miners, engineers, or of whatever talent would be needed. Money would not be an issue for the church as their assets were the faith and devotion of the people! Why would such a mission be sanctioned? I am certain any curious or skeptical mind will ask. They wanted to hide something that was not meant to be known. Something so powerful in the knowledge of its existence would be detrimental to some or could completely alter history in an unfavorable way in the minds of others. Why not hide it! Of all the suspects mentioned herein, and there are others not mentioned, only a couple known to exist in history could even be considered as wanting to leave a giant sacred symbol in a relatively virgin land to mark a site such as Nolan's Cross!

How can there be any assumption that the church has done what is found on Oak Island? There is some evidence left in history that gives rationale to the theory. First, let's go back in history to one of the earliest religious events known by many today, the hiding of the Dead Sea Scrolls. The Dead Sea Scrolls containing some 800 plus pieces of ancient parchment were just recently discovered in the Qumran Caves of the Dead Sea region of old Palestine. The scrolls were hidden there for fear of their destruction during the Roman occupation of the Holy Lands around 68 A.D. It has only been recently these writings have begun to be analyzed by the use of modern technology that promises to reveal what has been lost for over 2000 years! So far the content of the parchments has revealed early Jewish and Hebrew religious writings including those containing prayers, hymns, writings akin to the Old Testament as well as an early description of the Ten Commandments. As time and science unravels this discovery that contains knowledge and mores of the time, will there be new things learned about the time of Jesus?

What is relevant about this event is that history has left an example that some things, even those of a religious matter must be hidden to protect the knowledge contained. There is still hope that someone or some group really has hidden the Holy Grail and the Ark of the Covenant still containing the Ten Commandments!

It is very acceptable that religious artifacts were carefully hidden under Solomon's Temple to avoid discovery and for their protection during attacks on the temple by the enemies of Jerusalem! There are many books written where others claim there is much more to history than what is known about religious events surrounding Christianity! So they imply there is still some knowledge hidden that for now can only be speculation until proven otherwise. We must all try to be open-minded about these things, yet there are many who are not. There are still those who doubt space travel. Could there be a possibility that something of a religious significance is buried on Oak Island? Yes! Could the Catholic Church have done it? Maybe!

Here are more things to ponder about the church being involved in Nolan's Cross. Let's look back a few pages at some of the symbols included herein, beginning with the Amiens Rose. The importance of the Rose (it has religious meanings) is first that is found at one of the most important religious places in existence to the Catholic Church other than the Vatican City – Notre Dame Cathedral! The cathedral is located in Paris, France on an island in the Seine River and was built as a dedication to the Virgin Mary. The construction of the church took place from 1163 to its completion a hundred years later. This is also during the time of the rise of the Knights Templar who were probably involved in the construction of the cathedral along with some fellows from the Freemasons. It is important to remember that the Templar built many churches throughout Europe and in the Holy Lands in this era. What is of significance about this church is what is seen as the Amiens Rose! Looking back again at the geometry of the rose, it is a pentagram star tilted downward to the left at an angle exactly the same as Nolan's Cross and the symbol found by Sir Henry Lincoln. This is not a coincidence!

The other sacred symbol version of the tilted star was found at the Templar built church to honor St. Bartholomew who was one of the 12 apostles of Jesus. The church located near Soria, Spain was used as a refuge for the Knights Templar in the early 13th century, an area with a great Templar presence. The Templar had used a castle nearby as a defensive position to block enemies from approaching up the Rio Lobos Canyon. The symbol found on an outside wall is the downward tilted star encircled with what are known as hearts, a symbol for the sacred heart of Jesus!

These two churches both displaying the downward left-tilted star are dated around the 12th century and give another indication as to the significance of that symbol in its relevance to the sacred symbols found at Rennes and Oak Island!

Another part of the sacred symbol geometry at Rennes and Oak Island is the overlapping circles known as the vesica piscis. The importance of the circle alone has already been proven to be a sacred symbol for many reasons since the beginnings of humankind. The overlapping of the two circles has also been used since the earliest of times as a symbol representing the union of god and goddess united for creation. Like any other symbol, there are different interpretations by different users. Another view of the vesica piscis is of the vertical center intersection area seen as a mandrola, the elongated oval shape surrounding an image used in Christianity showing a religious being such as Christ or the Virgin Mary. The mandrola has been displayed in churches as an arch design in windows, entrance ways, and some construction beginning around 2 A.D. The design of the vesica piscis used in Christian art is believed to have been adopted from the geometric mathematics developed by Pythagoras. An often seen display of the mandrola shape is its two horizontal intersecting arches displayed as a fish symbol often seen today on windows and trunks of automobiles usually driven by Christians. The symbol of the fish also known as ichthus, a Greek acronym for

Jesus, is to many the very recognizable symbol referring to Jesus who showed the disciple Peter how to catch fish. There are other meanings besides these few that can easily be found in many resources. This important display of the two circles intersecting and creating the mandrola has for millenniums been a symbol used in Christianity with sacred meaning.

A couple very sizable displays of the vesica piscis can be found at the Washington Monument and at St. Peter's Square. The Washington Monument was built to honor the first president of the United States, George Washington. The construction of what is an obelisk was begun on July 4, 1848, taking four years to complete before opening to the public. The monument stands at 555 feet and was once the tallest structure in the world. Obelisks were first used in Egypt as symbolic entrances to sacred sites. Maybe the one in D.C. was meant to be an entrance to a special place as well. (This is only my opinion; yours may be different.) The monument is ringed with the vesica piscis with the base of the obelisk centered in the elongation of the mandrola! The Freemasons had a hand in the construction and dedication of the monument as well, as did George Washington who was a Freemason.

St. Peter's Square, located in Rome, is the site where the apostle St. Peter was crucified upside down on the cross dying as a martyr for Christianity in 64 A.D. It was the Roman Emperor Nero who ordered St. Peter's death. St. Peter is considered the First Pope by many and the early leader of the movement toward Christianity. The eventual success of the Catholic Church over the Roman Empire permitted the site of St. Peter's death and burial to eventually be recognized as sacred ground. In 1568, a 25-foot tall red granite obelisk arrived from Egypt that was estimated to be 4000 years old and was placed at its present location by the designer Domenico Fontana. A later designer, Lorenzo Bernini, in 1656 to 1667 created a long oval shaped site that is 240x320 meters that encompasses the obelisk with

The Washington Monument

St. Peter's Square

two opposing symmetrical arch colonnades in front of the famous Catholic Basillica. When the Pope appears on the balcony of the Basillica to address his followers, he is looking directly at St. Peter's Square where the masses congregate. The two opposing colonnades can be visualized as creating two perfectly symmetrical intersecting circles that create a vesica piscis with a mandrola at their intersection and the obelisk in its center.

All of the sacred symbols described in this book are all relating to one subject: humankinds importance in belief of the divine. The sacred symbol of the star or pentagram, especially in its orientation tilting downward, and the union of the vesica piscis found in the encrypted symbols of Rennes-le-Château and Oak Island have together carried a secret sacred message. Hopefully, you are formulating what it might be!

There is the other question, though, about who orchestrated what happened on Oak Island: was it the Knights Templar or the Catholic Church? Both wanted to conceal a secret. There is something hidden on Oak Island. Why? Was it done as protection from those who would exploit its existence or was it intended to never be exposed? Either way, maybe whatever it is had to be concealed for protection until the world could manage its discovery. What is it that has been hidden there so cleverly on Oak Island?

What is important for this book is the discovery of the sacred geometry symbol decoded in Nolan's Cross and its unique connection to the entangled story of Rennes-le-Château and earlier events in history. Could it be that whatever secret was discovered at Rennes by Sauneire was eventually secreted to Oak Island, just like the sacred symbol was designed in the cross to let the world know that what is sought is now there?

Where does it end? If something has been hidden on Oak Island, and it is what I think it could be, that something could answer one of the most holiest questions asked in all of Christianity! Whatever

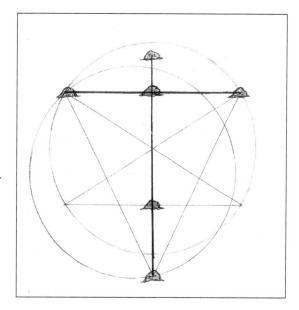

Nolan's Cross and the Sacred Symbol Illustration by the author

is there will surely add new chapters into the books not only about Oak Island, but about history itself. The History Channel TV series *The Curse of Oak Island* has been great mystery television for almost ten years. Now with my discovery of the sacred symbol encrypted in Nolan's Cross, I wonder what will happen next? Will the Lagina brothers and Tom Nolan, the person who owns the land where the cross sits, continue to look for a buried treasure in the Money Pit or in the swamp or will they all finally see the real treasure of Oak Island? There is a bigger treasure in the unraveling of the story than any cache of precious metals they can hold in their hands.

The answer lies in Nolan's Cross!

References

Books

Knights Templar in the New World by William Mann

Oak Island by William S. Crooker

Sacred Geometry: Your Personal Guide by Bernice Cockram

The Templar Mission to Oak Island and Beyond by Zena Halpern

Oak Island Secrets by Mark Finnan

The Curse of Oak Island by Randal Sullivan

The Knights Templar: God's Warriors, The Devil's Bankers by Frank Sanello

The History of the Knights Templar by Charles Addison

The Secret Treasure of Oak Island by D'Arcy O'Connor

Mary Magdalene: Women, the Church, and The Great Deception by Europa Compass

The Templars: The Rise and Spectacular Fall of God's Holy Warriors by Dan Jones

The Scrolls of Onteora: The Cremona Document by Donald Ruh

The Templars by Piers Paul Read

God's Wolf by Jeffrey Lee

Code Breaker: The History of Secret Communications by Stephen Pincock

The Freemasons: A History of the World's Most Powerful Secret Society by Jasper Ridley

Secret Societies by John Lawrence Reynolds

The Templars and the Assassins by James Wasserman

The Templars and the Ark of the Covenant by Graham Phillips

The Golden Age of Piracy: The Truth Behind Pirate Myths by Benerson Little

The Knights Templar by Helen Nicholson

The Hooked X by Scott F. Wolter

The Tempar Pirates: The Secret Alliance To Build the New Jerusalem by Ernesto Frers

Holy Grail Across the Atlantic by Michael Bradley

Holy Blood Holy Grail by Michael Baigent, Richard Leigh, Sir Henry Lincoln

The Templar Code for Dummies by Christopher Hodapp

The Lost Symbol by Dan Brown

The Oak Island Encyclopedia by Hammerson Peters

The Books Of Kings: (Biblical) by Chuck Missler & Rev. Dean DeFino

Oak Island And Its Lost Treasure by Graham Harris & Les MacPhie

The Templars Secret Island: The Knights, The Priest and The Treasure

by Erling Haagensen & Henry Lincoln

The Lost Treasure of the Knights Templar: Solving the Oak Island Mystery by Steven Sora

Key to the Sacred Pattern by Henry Lincoln

Sacred Geometry by Bernice Cockram

The Accursed Treasure of Rennes-le-Chateau by Gerard de Sede & translated by Sir Henry Lincoln

Oak Island's Mysterious "Money Pit" by David MacDonald in *Reader's Digest,* January 1965

The Holy Place: Saunière and the Decoding of the Mystery of Rennes-le-Chateau by Henry Lincoln

Freemasonry: Symbols, Secrets, Significance by W. Kirk MacNulty

Other

Magen David Symbol letter, dated April 5, 2008. Ohr Somayach Institute. www.ohr.edu. Written by Rabbi Yirmiyahu Ullman

On a property of the vesica piscis. Mauro Maria Baldi, Department of Computer Science at the University of Milano-Bicocca, Milan, Italy.

The Holy Bible. Old and New Testaments. American Bible Society, 1816.

The Curse of Oak Island. Seasons 1-9. History Channel / Prometheus Entertainment.

Websites

Wikipedia was a quick search tool that I often used as an initial starting point. Other websites that I used include but are not limited to the following:

www.britishexplorers.com/woodbury/raleigh1.html

htpps://www.thoughton.com/captain-william-kidd-2136225

www.thepirateking.com/bios/easton_peter.htm

www.freemasons-freemasonary.com/pentagram_freemasonary.html

www.thepirateking.com/bios/index.htm

https://archive.macleans.ca/article/1950/9/15/newfoundlands-pirate-king

https://templehousegallery.com/nauvo-temple-star-window-history

https://www.jewishvirtuallibrary.org/magen-david

https://www.aish.com/star of david/

https://www.bh.org.il/blog-items/star-of-david-birth-of-a-symbol-legends-vs-facts/

https://www.newworldencyclopedia.org/entry/Pentagram

https://western-hindu.org/2009/10/20/shaivite-hindu-symbols-the-shatkona/

https://www.gotquestions.org/pentagram-meaning.html

https://www.museumreplicas.com/knights-templar-symbols#text

https://templehousegallery.com/nauvo-temple-star-window-history

https://www.mysteriesofcanada.com/wp-uploads/2016/11/nolans-cross.jpg

Themysteryofoakisland.weekly.com/uploads/2/8/2/4/28244355/216275528.jpg?269-Amundson-tree-of-life-over-nolans-cross

https://www.rmg.co.uk/discover/sir-John-Hawkins

www.piratesurgeon.com/pages/surgeon_pages/quarters2.html

www.thewayofthepirates.com/famous-pirates/peter-easton/

https://www.thoughton.com/captain-william-kidd-2136225

www.britishexplorers.com/woodbury/raleigh1.html

https://www.britianica.com/biography/William-Kidd

https://www.novascotia.com/see-do/attractions/port-royal-national-historic-

https://www.canadiangeographic.ca/article/tribute-chester-nova-scotia

https://www.thing.de/artwarpeace/1consul/sebstry2.htm-seborga

https://www.sanctisepulchrie.org/history//lang=en-castrum-sepulchri

https://www.dualcroaaroads.com/post/the-histyory-and-symbolisim-of-the-pentagram

https://www.ancient.eu/Tyre/

https://www.onetribeapparel.com/blogspai/meaning-of-tree-of-life

https://kris10spredictions.com/tag/the-evans-stone/

https://www.smithsonian.com/history/story-behind-forgotten-symbol-american-revolution-liberty-tree-180959162/

https://www.gotquestions.org/cedars-of-Lebanon-html

https://www.revoluntionary-war-and-beyond.com/washington-crusiers-flag.html

https://www.reddit.com/r/OakIsland/comments/blog3i5_is_this_the_biggest

casado.net/seborga/history/index.html

https://concen.org/node/37699/secret (viking-sea-chart-rosslyn)

https://crusadehistory.wordpress.com/2019/11/23/priory-of-sion-rrene-danjou/

https://crusadehistory.wordpress.com2019/07/13/French-templar-escapees/

https://crusadehistory.wordpress.com/2016/08/15/the-st-clairs-of-roslin/

https://www.theknightstemplar.org/twelve-french-templars-who-escaped

www.jamesonfamily.org/Sinclair_Gunn.html

clangunn.weebly.com/on-a-gunn-helping-discover-north-america-sir-james

esotericroslyn.org/templar-and-other-symbols-in-roslyn.html

www.grahmphillips.net/ark/ark5.html

https://ashleycowle.com/secret-viking-sea-chart/

www.unexplainedstuff.com/Objects-of-mystery-(ark of the covenant)

https://www.ancient.eu/Knights_Templar/

http://www.sacredmysterytours.com/(knights)

https://hubpages.com/education/The-Templar-Fleet-of-La-Rochelle

https://www.ancient.eu/Knights_Templar/(Ancient History Encyclopedia)

www.jasoncolavito.com/blog/nova-scotias-sinclair-castle

www.grahmphillips.net/ark5.html(Ralph de Sudeley)

www.jasoncolavito.com/the-templars-the-holy-grail-henry-sinclair.html

www.criticalenquiry.org/oakisland/masonic.shml

www.treasurenet.com/forums/treasurre-legends/472035-tunnels-more-just-found

https://historicnovascotia.ca/items/show/156(LaHave)

symboldictionary.net/?p=378(star over archway Jerusalem)

www.gjbath.com/OakIlsand/OI3EE.html(Nolan"s Cross & Tree of Life)

https://www.gaia.com/article/tree-of-life-sacred-geometry

https://wisdomfromnorth.com/the-shakespeare-mystery-ft-peter-amundsen-i2-3/

https://www.britanica.com/topic/Star-of-David

https://blog.judaicawebstore.com/the-star-of-david-a-brief-history

https://mfa.gov.il/MFA/MFA-Archive/1999/Pages/King-Solomon

https://www.khanacademy.org/humanities/world-history/ancient-medieval/early-

https://www.ancient.eu/King_David

Ancient-Symbols.com

https://www.britanica.combiography/David

https://www.geometriasacra.com/en/sacred-geometry-php(tree of life)

https://christinansymbolsinart.wordpress.com/2014/11/26/triangle/

https://www.todayifoundout.com/index.php/2011/09/origins-of-the-jolly-roger/

https://time.com/5877435/freemason-secrecy/

www.themasonictrowel.com/Articles/degrees/degree_3rd_files/the_five_points

https://www.universeofsymbolism.com/tree-symbolisim.html

https://www.bontanical-onl;ine.com/en/botany/holm-oak

https://famous-mathematicians.org/luca-pacioli/

https://rct.uk/collection/912660/studies-of-water-Leonardo

www.yourguidetoitaly.com/famous-italian-explorers.html

https://francisbaconsociety.co.uk/shakespeare-authorship/bacon-shakespeare

https://www.biography.com/explorer/francis-drake   https://bbcamerica.com/anglophenia/2011/10/did-shakespeare-really-write

https://rosicruciansinportlandoregonwilsonville.com/2019/05/13/rosicrucians-

https://oakislandlotfive.com/robert-s-young

https://www.oakislandtours.ca/early-history.html

https://www.biography.com/explorer/rene-robert-cavelier-sieur-de-la-salle

https://www.google.com/search?q+knights+templar-battle+flag&tbm

https://www.tempelherreden.org/templar-battle-flag

https://www.phoenician.org/solomons_temple.html

https://www.dailymail.co.uk/sciencetech/article-2536549/King-Solomon

https://www.neverthristy.org/bible/qa/qa-archives/question/who-helped-solomon

https://www.booksfact.com/religious/jewish-star-david-symbol-vedic-anahata

https://www.ancientpages.com/2018/02/12-masonic-symbols-explained

https://facts.net/types-of-crosses/

https://christiananswers.net/q-abr/abr-a013.html-crosses

https://howtodiscuss.com/t/cross-proportions/70167

htpps://www.pinterest.com/pin/399553798190270294/proportions-cross

https://symbolikon.com/downloads/tree-of-life-sephiroth-sacred-geometry/

https://templehousegallery.com/nauvo-temple-star-window-history

https://www.museumreplica.com/knights-templar-symbol:~:text=The Beauceant

https://www.newworldencyclopedia.org/entry/Pentagram

Elizabethan.org/compendium/6.html-pound-coin

https://www.learntotrade.co.uk/history-of-the-pound/

https://roslandcapital.com/products/sovereign-gold-coin

https://www.goldenarttreasures.com/article.php?id+4367-gold-sovereign

www.templiers.org/bezu-eng-php

https://www.catharcastles.info/montsegur.php

https://www.geni.com/people/Sir-Henry-of-Roslin-7th-lord-of-Roslin/#4

https://www.geni.com/people/Sir-William-Sinclair-of-Roslin/#5

https://www.gen.com/people/William-Sinclair-of-Roslin/#6

www.krausehouse.ca/krause/FortressOfLouisbourgResearcWebSearch/Ducdanville

www.grahmphillips.net/ark/ark6.html/Herdewyke

https://www.geni.com/people/Robert-de-St-Clair#2

https://www.nps.gov/jame/learn/historyculture/a-short-history-of-jamestown.html

www.loc.gov/teachers/classroommaterials/presentation/colonial/jamestown

Httpds://www.nps.gov/jame/learn/historyculture/prelude-to-Jamestown.htm

https://www.history.com/news/what-happened-to-the-lost-colony-of-roanoke

https://wwwoakislandcompendium.ca/blockhouse-blog/on-the-trail-vaughn

https://www.oakislandcompendium.ca/blockhouse-blog/daniel-mcginnis

https://ancient.eu/article/897/the-phoenicians-mariners/

www.sacredmysterytours.com/sacred-geometry#sthash

www.templars.org/bezu-eng.php

www.cnn.com/2013/02/28/world/americas/phoenician-colombus-america-sailboat.

https://www.theguardian.com/science/1999/nov/28/archaeology.uknews/sailors

https://www.ancient.eu/Greek_Alphabet/phoenician

http://www.phoenician.org/sea_peoples.htm-tyre

http://www,phoenician.org/cedars_of_lebanon.htm-acacia-tree

http://www.phoenician.org/ancient_ships.htm

http://www.iro.umontreal.ca/~vaucher/History/Ships_Discovery/northern-cog

https://www.sanctisepulchri.org/history/?lang=en-seborga

www.templars.org/bezu-eng.php

www.sacredmysterytours.com/sacred-geometry#sthash

http://www.henrylincoln.co.uk/geometry.php

https://sites.google.com/site/renneslechateauforbeginners/research

http://mathworld.wolfarm.com/pentagram.html

About the Author

LEE LARIMORE is a former math teacher who became interested in watching *The Curse of Oak Island* TV series that was first aired on January 5, 2014 by the History Channel. Ever since then, Lee has been captured by the mystery haunting the island for more than 220 years. Having already been intrigued by other legends like the Freemasons, pirates, lost treasures, and ancient historical mysteries, Oak Island became a must-not miss episode. With his background in math and a natural desire to solve things, he began to do research into the history of the potential suspects who could hold the answer to the mystery. Little did he know there really was an answer, but it was not what everyone thought and this is why he wrote this book!

Lightning Source UK Ltd.
Milton Keynes UK
UKHW022210051222
413407UK00015B/2889